ASCE Manuals and Reports on Engineering Practice No. 144

Hazard-Resilient Infrastructure

Analysis and Design

Sponsored by the
Infrastructure Resilience Division of ASCE

Edited by
Bilal M. Ayyub, Ph.D., P.E.

ASCE AMERICAN SOCIETY OF CIVIL ENGINEERS

Published by the American Society of Civil Engineers

Library of Congress Cataloging-in-Publication Data

Names: Ayyub, Bilal M., editor. | American Society of Civil Engineers. Infrastructure Resilience Division.
Title: Hazard-resilient infrastructure : analysis and design / sponsored by the Infrastructure Resilience Division of ASCE ; edited by Bilal M. Ayyub, Ph.D., P.E.
Description: Reston, Virginia : American Society of Civil Engineers, [2021] | Series: ASCE manuals and reports on engineering practice ; no. 144 | Includes bibliographical references and index. | Summary: "MOP 144 provides guidance and underlying framework for creating consistency across hazards, systems, and sectors in the design of new infrastructure systems and in enhancing the resilience of existing ones"-- Provided by publisher.
Identifiers: LCCN 2021005960 | ISBN 9780784415429 (print) | ISBN 9780784415757 (paperback) | ISBN 9780784482537 (ebook) | ISBN 9780784483459 (ePub)
Subjects: LCSH: Structural design. | Structural analysis (Engineering) | Fault tolerance (Engineering) | Building failures--Prevention. | Public works--Protection. | Disasters.
Classification: LCC TA658.4 .H39 2021 | DDC 624.1/7--dc23
LC record available at https://lccn.loc.gov/2021005960

Published by American Society of Civil Engineers
1801 Alexander Bell Drive
Reston, Virginia 20191-4382
www.asce.org/bookstore | ascelibrary.org

***Errata:* Errata, if any, can be found at https://doi.org/10.1061/9780784415757.**

ISBN 978-0-7844-1575-7 (print)
ISBN 978-0-7844-8253-7 (PDF)
ISBN 978-0-7844-8345-9 (ePub)

Manufactured in the United States of America.

27 26 25 24 23 22 21 1 2 3 4 5

MANUALS AND REPORTS ON ENGINEERING PRACTICE

(As developed by the ASCE Technical Procedures Committee, July 1930, and revised March 1935, February 1962, and April 1982)

A manual or report in this series consists of an orderly presentation of facts on a particular subject, supplemented by an analysis of limitations and applications of these facts. It contains information useful to an average engineer in his or her everyday work, rather than findings that may be useful only occasionally or rarely. It is not in any sense a "standard," however, nor is it so elementary or so conclusive as to provide a "rule of thumb" for nonengineers.

Furthermore, material in this series, in distinction from a paper (which expresses only one person's observations or opinions), is the work of a committee or group selected to assemble and express information on a specific topic. As often as practicable, the committee is under the direction of one or more of the Technical Divisions and Councils, and the product evolved has been subjected to review by the Executive Committee of the Division or Council. As a step in the process of this review, the proposed manuscripts are often brought before the members of the Technical Divisions and Councils for comment, which may serve as the basis for improvement. When published, each manual shows the names of the committees by which it was compiled and indicates clearly the several processes through which it has passed in review, so that its merits may be definitely understood.

In February 1962 (and revised in April 1982), the Board of Direction voted to establish a series titled "Manuals and Reports on Engineering Practice" to include the manuals published and authorized to date, future Manuals of Professional Practice, and Reports on Engineering Practice. All such manual or report material of the Society would have been refereed in a manner approved by the Board Committee on Publications and would be bound, with applicable discussion, in books similar to past manuals. Numbering would be consecutive and would be a continuation of present manual numbers. In some cases of joint committee reports, bypassing of journal publications may be authorized.

A list of available Manuals of Practice can be found at http://www.asce.org/bookstore.

DEDICATION

Richard N. Wright, Ph.D., P.E., NAE, Dist.M.ASCE
May 17, 1932 to May 31, 2019

Dr. Wright was a research professor of civil and environmental engineering at the University of Maryland, College Park, the past chair of ASCE's Committee on Adaptation to a Changing Climate, a member of the ExCom of the Infrastructure Resilience Division, and a member of the ASCE Committee on Sustainability, as well as the chair of its Sustainable Infrastructure Education Subcommittee. He was a distinguished member of ASCE and a member of the National Academy of Engineering.

He received his bachelor's and master's degrees from Syracuse University and his doctorate from the University of Illinois at Urbana-Champaign, all in civil engineering. He worked as the director of the Building and Fire Research Laboratory of the National Institute of Standards and Technology (NIST) and a professor of civil engineering at the University of Illinois at Urbana-Champaign. He was the president of the International Council for Research and Innovation in Building and Construction (CIB) and the president of the Liaison Committee of International Civil Engineering Organizations.

Dr. Wright's professional honors include the Gold Medal of the Department of Commerce (1982); Federal Engineer of the Year from the National Society of Professional Engineers (1988); Mahaffey Award from the National Conference of States on Building Codes and Standards (1998); Michel Award for Industry Advancement of Research from the Civil

Engineering Research Foundation (1999); election to the National Academy of Engineering (2003); International Award of the Japan Society of Civil Engineers (2003); University of Illinois, College of Engineering, Alumni Award for Distinguished Service (2006); President Medal of ASCE (2010); and the Hans Oser Standards Alumni Association Distinguished Service Award (2014).

CONTENTS

FOREWORD

ASCE has established the Grand Challenge as a way of recognizing the need for increasing investment in the infrastructure on which our society is built. The systems on which we rely to provide safe transportation, clean water, secure structures, and reliable energy and telecommunications are increasingly frail as they reach the end of their design lives.

However, it is not enough to simply replace old with similar. Limitations on natural and fiscal resources, a changing climate, and expectations of modern infrastructure to have a low life-cycle cost and increased longevity are making the ways of the past obsolete. Today's civil engineers need to consider new design and construction approaches incorporating new materials and emerging technologies. We must ensure that the next generation of infrastructure systems not only meets nominal expectations but also has the resilience to perform during and recover rapidly after exposure to hazardous conditions.

Recognizing this, ASCE has established roughly two dozen policy statements (PS) related to infrastructure development and disaster mitigation. PS 500, *Resilient Infrastructure Initiatives*, was adopted to

> … support initiatives that increase resilience of infrastructure against man-made and natural hazards through education, research, planning, design, construction, operation and maintenance. Development of performance criteria and uniform national standards that address interdependencies and establish minimum performance goals for infrastructure is imperative.

This policy goes on to note that

> … an all-hazard, comprehensive risk assessment that considers event likelihood and consequence, encourages mitigation strategies,

monitors outcomes, and addresses recovery and return to service should be routinely included in the planning/design process for infrastructure at all government levels.

Similarly, PS 518, *Unified Definitions for Critical Infrastructure Resilience*, establishes consistent terminology to allow a clear characterization of critical infrastructure, hazards, and resilience. In this vocabulary, resilience refers to

> … the capability to mitigate against significant all-hazards risks and incidents and to expeditiously recover and reconstitute critical services with minimum damage to public safety and health, the economy, and national security.

Subsequent to this, and in continued response to the need to promote programs leading to design, construction, and maintenance of robust infrastructure systems, ASCE formed the Infrastructure Resilience Division (IRD) under the Committee for Technical Advancement (CTA). The charge for IRD is to develop "… products and services to include … manuals of practice … to advance resilient practices related to civil infrastructure and lifeline systems …".

After a significant investment of time and resources by ASCE members working through IRD, this document was developed, reviewed, and finalized. We feel you will find that the publication is noteworthy in the depth and breadth of its address of the topics of infrastructure resilience. It is the hope of IRD, CTA, and ASCE that you and others will find the Manual of Practice (MOP) to be an effective reference for engineers and others interested in the diversity of issues related to the planning, design, analysis, and construction of hazard-resilient infrastructure systems.

One will recognize that this MOP is significant in content which, in turn, may limit immediate application of the concepts and approaches included herein by those in small organizations or who have had less exposure to the contemporary issues facing the nation's infrastructure replacement needs. However, the authors submit that to make it less comprehensive would be an injustice to the topic and to those wishing to include robust system design in their practice. Further, with time, the utility of this document is expected to increase and the information contained herein become a standard of practice.

Before concluding, one must acknowledge that the creation of an MOP requires an immense effort by a cadre of volunteers. This team hopes of providing the user with an important reference resource to address 21st century system challenges. It is also worth noting that whereas many contributed significantly to the production of this document, a substantial effort in making the MOP a reality was invested by Dr. Bilal Ayyub, P.E.,

Dist.M.ASCE, Hon.M.ASME. Dr. Ayyub is a professor of civil and environmental engineering and director of the Center for Technology and Systems Management at the University of Maryland. Many involved with the project have suggested that, in no small way, his vision, leadership, and commitment to this project is a major factor that resulted in the completion of this MOP. It is the dedication to sharing best practices of our craft by volunteers like those involved in this project that allows ASCE to stand globally at the front of our profession.

Dennis D. Truax, Ph.D., P.E., DEE, D.WRE, F.ASCE, F.NSPE
ASCE President 2022
James T. White Endowed Chair, Department Head, and Professor
Department of Civil and Environmental Engineering
Mississippi State University
Director, Mississippi Transportation Research Center

PREFACE

This publication titled, *Hazard-Resilient Infrastructure: Analysis and Design*, was prepared by the ASCE Infrastructure Resilience Division during 2018 to 2020.

Benefits of enhancing resilience go well beyond cost savings and include important considerations such as the health, safety, and welfare of the public and socioeconomic well-being. Resilient systems should be developed to meet sustainability requirements defined by the three pillars of sustainability of environmental, social, and economic, by reconciling environmental, social equity, and economic demands, respectively. This MOP has primary objectives of providing guidance for the design of hazard-resilient infrastructure systems and enhancing the resilience of existing ones for the purpose of meeting community resilience goals. It secondarily contributes to the development and enhancement of standards for infrastructure analysis and design, as well as regulations and building codes that refer to them, in a world in which risk profiles are changing. This MOP and such efforts contribute to achieving hazard-resilient infrastructure.

This MOP offers a framework for creating consistency across hazards, systems and sectors. It offers an opportunity to prepare a series of hazard-specific MOPs on topics such as (a) earthquake, (b) flooding, (c) tornado and extreme wind, (d) fire, (e) climate, and others. MOP 140, *Climate-Resilient Infrastructure* already exists, prepared by the ASCE Committee of the Adaptation to a Changing Climate. This MOP could be used as a basis to prepare earthquake-resilient infrastructure MOP. In addition, it offers an opportunity to prepare a series of multi-hazard sector or system specific MOPs on topics such as (a) electric power distribution, (b) transportation (e.g., mass transit networks), (c) water distribution, (d) communication,

(e) marine transport and navigation, and others. For example, this MOP could be used as a basis to prepare a publication on resilient-electric power distribution infrastructure. Combinations of these two types are worthy of pursuits, such as an earthquake resilient-electric power distribution infrastructure MOP.

ACKNOWLEDGMENTS

The ASCE Infrastructure Resilience Division acknowledges the contributions of the Blue-Ribbon Panel Reviewers and in particular the suggestions provided by Mr. Glenn R. Bell, Professor Gerald E. Galloway, and Professor Armen Der Kiureghian. The committee also acknowledges the encouragement and the support provided by Mr. John E. Durrant, P.E., F.ASCE, ASCE's senior managing director for Engineering and Lifelong Learning, and the Committee on Technical Advancement, and Catherine Tehan, ASCE IRD director.

The team also acknowledges the commitment of ASCE to gender and racial equity.

Development Team

Editor
Bilal M. Ayyub, Ph.D., P.E., Dist.M.ASCE

Lead Authors
Bilal M. Ayyub, Ph.D., P.E., Dist.M.ASCE
Dave Butry, Ph.D.
Craig A. Davis, Ph.D., P.E.
Sanjeev R. Malushte, Ph.D., S.E., P.E. (Civil), P.E. (Mechanical), C.Eng, F.ASCE
Ricardo A. Medina, Ph.D., P.E., M.ASCE
Mahmoud Reda Taha, Ph.D., P.E., M.ASCE
Van de Lindt, Ph.D., F.ASCE

Contributing Authors
Cory R. Brett, P.E., M.ASCE
Sherif Daghash, Ph.D., P.E., M.ASCE
Caroline Field, C.Eng
Juan Fung, Ph.D.
Paolo Gardoni, Ph.D.

Sue McNeil, Ph.D., M.ASCE
Fernando Moreu, Ph.D., P.E., M.ASCE
Ali Mostatavi, Ph.D.
Yalda Saadat
Neetesh Sharma
Kenichi Soga, Ph.D., M.ASCE
Eslam Soliman, Ph.D., M.ASCE
Elaina J. Sutley, Ph.D., A.M.ASCE
Armin Tabandeh
Douglas Thomas
Eric Vugrin, Ph.D.
Richard N. Wright (deceased), Ph.D., P.E., NAE, Dist.M.ASCE

Advisory Committees
ASCE Infrastructure Resilience Division (IRD) ExCom
ASCE Committee on Technical Advancement (CTA)

Blue Ribbon Panel

Nominations Approved by the IRD ExCom
Glenn R. Bell, P.E., formerly with Simpson Gumpertz & Heger Inc.
Gerald E. Galloway, Ph.D., NAE, Dist.M.ASCE, University of Maryland, College Park
Armen Der Kiureghian, Ph.D., F.ASCE, University of California, Berkeley

Review Coordinators
Kent Yu, Ph.D., P.E., S.E., M.ASCE, *IRD ExCom Chair*
Catherine Tehan, Aff.M.ASCE, *ASCE Staff Contact, IRD Director*

Administrative Committees

Initiated by the 2018 ASCE-IRD Executive Committee
Kent Yu, Ph.D., P.E., S.E., M.ASCE, *Chair (2019–2020)*
John Willem Van De Lindt, Ph.D., F.ASCE, *Vice Chair (2019–2020)*
Bilal M. Ayyub, Ph.D., P.E., Dist.M.ASCE, *Past Chair (2017–2018)*
Nasim Uddin, Ph.D., P.E., F.ASCE, *Member*
Mahmoud Reda Taha, Ph.D., *Member*

Advisor
Craig A Davis, Ph.D., G.E., P.E., M.ASCE, *Past Chair*

Approval of Initiative by ASCE Committee on Technical Advancement (CTA)
CTA represented by
Jonathan Esslinger, M.ASCE, *ASCE Staff, Director of CTA*
Scott Murrell, P.E., M.ASCE, *CTA Liaison*
Catherine Tehan, Aff.M.ASCE, *ASCE Staff Contact, Director of IRD*

ACRONYMS

ABM	Agent-based modeling
AI	Artificial intelligence
AM	Additive manufacturing
AMoC	Additive manufacturing of concrete
ANN	Artificial neural networks
AR	Augmented reality
ASCE/SEI 24	*Flood Resistant Design and Construction*
ASCE/SEI 7	*Minimum Design Loads for Buildings and Other Structures*
BEIS	UK Department for Business, Energy and Industrial Strategy
BFE	Base flood elevation
BIM	Building information modeling
BMS	Building management system
BNTs	Boron nanotubes
BRA	Boston Redevelopment Authority
BRP	Blue Ribbon Panel
BS	British Standard
CC	Contour crafting
CGE	Computable general equilibrium
CIS	Civil infrastructure systems
CNC	Computer numerical controlled
CNT	Complex network theory
CoE	Center of Excellence (refers to the NIST Community Resilience Center of Excellence)
CPTED	Crime prevention through environmental design
CTA	Committee on Technical Advancement
DBPR	Department of Buildings and Professional Regulations
DEFRA	UK Department for Environment, Food & Rural Affairs
DFE	Depth of flood elevation

DFOSs	Distributed fiber optic sensors
DIC	Digital image correlation
DL	Deep learning
DOS	Number of days of supply
DOT	Department of Transportation
EBMH	Economic burden model of hazards
ECC	Engineered cementitious composites
FBG	Fiber Bragg grating
FEMA	Federal Emergency Management Agency
FHWA	Federal Highway Administration
FIRM	Flood insurance rate map
FIS	Flood insurance study
FRP	Fiber-reinforced polymers
GA	Genetic algorithms
GNPs	Graphene nanoplatelets
GVA	Gross value added
HACT	Housing Associations' Charitable Trust
HSHDC	High-strength high-ductility concrete
IA	Immediate action(s)
ICT	Instant communication technologies
IGBC	Interagency Green Building Committee
IO	Input–output
IOT	Internet of things
IRAM	Infrastructure resilience analysis method
IRC	Intelligent reinforced concrete
IRD	Infrastructure Resilience Division
IRR	Internal rate of return
LCC	Life-cycle cost
LiDAR	Light detection and ranging
LoS	Level of service
MassDOT	Massachusetts Department of Transportation
MCE	Maximum credible earthquake
MCEER	Multidisciplinary Center for Earthquake Engineering Research
MHHW	Mean-higher-high-water
ML	Machine learning
MLGW	Memphis Light, Gas, and Water
MOP	Manual of Practice
MSL	Mean sea level
MTL	Mean tide level
MTPD	Maximum tolerable period of disruption
MWCNTs	Multi-walled carbon nanotubes
NAVD88	North America Vertical Datum of 1988

NEI	Nuclear Energy Institute
NIST	National Institute of Standards and Technology
NiTi	Nickel–titanium
NMSZ	New Madrid Seismic Zone
NOAA	National Oceanic and Atmospheric Administration
NO_x	Nitrogen oxides
NPV	Net present value
NRC	US Nuclear Regulatory Commission
NWS	National Weather Service
PC	Polymer concrete
PDD	Presidential Policy Directive
PE	Polyethylene
PM	Particulate matter
PV	Present value
RC	Reinforced concrete
RDM	Robust decision making
RE	Recovery effort
REDi	Resilience-based Earthquake Design Initiative
RTO	Recovery time objectives
SAM	Social accounting matrix
SAR	Synthetic aperture radar
SFM	Structure from motion
SHCC	Strain hardening cement composites
SHM	Structural health monitoring
SI	Systemic impact
SIFT	Scale-invariant feature transform
SIR	Savings-to-investment ratio
SLC	Sea-level change
SLOSH	Sea, Lake, and Overland Surges from Hurricanes
SLR	Sea-level rise
SMA	Shape memory alloy
SOP	Standard operating procedure(s)
SoS	System of systems
SO_x	Sulfur oxides
SP	Actual system performance level
sqkm	Square kilometer
TPV	Total project value
TRE	Total recovery effort
TSP	Targeted system performance level
TVA	Tennessee Valley Authority
UAS	Unmanned aerial systems
UAV	Unmanned aerial vehicles
UHP	Ultra-high performance

USACE	United States Army Corps of Engineers
VO	Value of options
VR	Virtual reality
VUCA	Volatile, uncertain, complex, and ambiguous
WSN	Wireless sensor network

EXECUTIVE SUMMARY

According to the United Nations Office for Disaster Risk Reduction (UNDRR), by 2025 half of the world's inhabitants are expected to increase by roughly two-thirds, and the vast majority of property and wealth is concentrated in urban centers situated in locations already prone to major disasters, such as earthquakes and severe droughts, and along the flood-prone coastlines (UNDRR 2012). The *Sendai Declaration*, among others, called for the strengthening of disaster risk reduction to mitigate loss of lives and assets worldwide (UNDRR 2015), in other words, a call to strengthen critical infrastructure. Infrastructure systems are essential for supporting communities, changing the equity imbalance, and increasing resilience (United Nations 2016). It is therefore essential to develop means for increasing the infrastructure system resilience against severe hazards.

Civil infrastructure systems traditionally have been designed, constructed, operated, and maintained for appropriate probabilities of functionality, durability, and safety while exposed to extremes during their full-service lives. Examining systems in the context of resilience would add proper considerations for adaptability to changing conditions including recovery, not to mention decision making about improved resilience of existing infrastructure.

The purpose of this MOP is to provide guidance for and contribute to the development or enhancement of standards for hazard-resilient infrastructure. The goal is to transform our infrastructure and make it hazard-resilient. The framework provided here emphasizes infrastructure systems and how they support community resilience. Any infrastructure system is a combination of components or systems constructed over time. Therefore, analysis of existing systems is just as relevant, if not more so, than new.

The underlying approaches in this publication are based on using probabilistic methods for risk analysis and management for infrastructure projects to effectively address uncertainties within a planning horizon timeframe. The approach provided herein focuses on identifying and analyzing hazards, system failure, associated probabilities, and consequences including direct and indirect losses, failure and recovery profiles, quantification of resilience, impacts on communities, economics of resilience, and technologies for enhancing resilience for new as well as the existing infrastructure.

The first chapter sets the context, articulates the needs, provides an objective and the scope of work statement for the manual, lists topics warranting additional analysis, and defines users and the uses of the manual. It also provides perspectives on the relationships between infrastructure resilience and its sustainability. Resilience and sustainability are related concepts, and having an understanding of this relationship would enhance designs and performances of infrastructure systems. The chapters that follow achieve the set goal by covering their respective areas. Examples and cases studies are provided in all subsequent chapters to meet the needs of practitioners.

Chapter 2 provides a methodological framework for achieving hazard-resilient infrastructure. The framework is designed to be generic and broad in terms of applicability and offers a road map for using the materials provided in subsequent chapters.

Chapter 3 describes available and mature methods for assessing the resilience of systems and facilities individually and collectively as systems of systems and network. This core chapter provides a primary building block in the engineering phase and other phases.

Chapter 4 focuses on the economics of resilience and risk management. It summarizes methods used by microeconomists for examining resilience-enhancing alternatives to utilize resources most effectively.

Chapter 5 provides an overall design approach of resilient-infrastructure systems. The chapter uses several examples and case studies to illustrate concepts introduced.

Chapter 6 covers community socioeconomics and offers guidance on ways to account for such an important consideration in design or retrofitting existing infrastructure systems. It also introduces examples to illustrate such concepts.

Chapter 7 provides a review of emerging resilience-enabling technologies for new and existing infrastructure systems, such as smart materials, advanced construction technology, advanced sensing technology, and so on. It also provides insights in assessing the maturity of such technologies.

This MOP includes an appendix on terminology.

REFERENCES

UNDRR (United Nations Office for Disaster Risk Reduction). 2012. *Making cities resilient: My city is getting ready! A global snapshot of how local governments reduce disaster risk*. Geneva: United Nations. Accessed March 8, 2021. https://www.unisdr.org/campaign/resilientcities/home/article/moving-to-the-decade-of-action-with-mcr2030.

UNDRR. 2015. *Sendai framework for disaster risk reduction 2015–2030*. 1st ed. Geneva: United Nations. Accessed April 7, 2019. https://www.unisdr.org/files/43291_sendaiframeworkfordrren.pdf.

United Nations. 2016. "The infrastructure—inequality—resilience Nexus." In *The global sustainability development report*, chapter 2. New York: Dept. of Economic and Social Affairs.

DISCLAIMER

Members of the ASCE Infrastructure Resilience Division (hereafter referred to as ASCE IRD) prepared this document titled, *Hazard-Resilient Infrastructure: Analysis and Design*. Although this product was produced using the best available resources, ASCE IRD makes no warranty, expressed or implied, or assumes any legal liability or responsibility for the accuracy, completeness, or usefulness of any information, apparatus, product, or process disclosed, or represents that its uses would not infringe privately owned rights. Reference herein to any specific commercial product, process, or service by trade name, trademark, manufacturer, or otherwise does not necessarily constitute or imply its endorsement, recommendation, or favoring by the ASCE IRD.

Although this publication provides extensive references, it is not intended to include a comprehensive literature review of all the topics covered nor the work of other meritorious experts. Such a comprehensive review is beyond the scope of this manual, which is intended to inform engineering practice by bringing together a survey of various approaches currently available to the professional engineer. The reader is expected to utilize his or her expert judgment to determine the applicability of any particular method to any challenge at hand.

CHAPTER 1
INTRODUCTION

1.1 NEEDS AND SIGNIFICANCE

The United Nations Office for Disaster Risk Reduction (UNDRR 2012) reported that by 2025 half of the world's inhabitants are expected to increase to roughly two-thirds of its population, and the vast majority of property and wealth is concentrated in urban centers situated in locations already prone to major disasters, such as earthquakes and severe droughts, and along the flood-prone coastlines. It also reported that the 2011 natural disasters resulted in direct damages of $366 billion and 29,782 fatalities worldwide. The Third UN World Conference on disaster risk reduction was held in Sendai City, Miyagi Prefecture, Japan, in March 2015. The *Sendai Declaration*, among other things, called for the strengthening of disaster risk reduction to mitigate loss of lives and assets worldwide (UNDRR 2015). In 2011, storms and floods accounted for up to 70 of 302 natural disasters worldwide, with earthquakes producing the greatest number of fatalities (UNDRR 2012). As an example, the 2012 annual loss in the United States as a result of all hazards is about $55 billion. It is anticipated that such disasters would occur in increasing trends of storm rates and disaster impacts because of a combined effect of climate change and increased coastal inventory of assets (Ayyub et al. 2012).

Infrastructure systems are essential for supporting communities, changing the equity imbalance, and increasing resilience (United Nations 2016, NRC 2012). It is therefore essential to develop means for increasing the infrastructure system resilience against severe hazards. Although no population center or a geographic area can be risk free from natural or human-caused hazards, communities should strive to enhance resilience to the destructive forces or the impacts of events that may claim lives and damage property. Gilbert (2010) provided population- and wealth-adjusted loss and fatality count trends from 1960 to 2009 to demonstrate that both are about flat without significant slopes; however, the United States is becoming more vulnerable to disaster because of increased population concentration in areas prone to natural disasters (Burby 1998, Berke et al. 1993), and not only is our infrastructure in a persistently inadequate condition but also it continues to deteriorate (ASCE 2009).

Enhancing system resilience at the infrastructure, network, community, and so on levels could lead to massive savings through risk reduction and expeditious recovery. The rational management of such reduction and recovery is facilitated by an appropriate definition of resilience and associated metrics. Current definitions do not always lend themselves naturally and intuitively for the development of consistent resilience metrics. These metrics should have clear relationships to the metrics of the relevant abstract notions of reliability and risk. Definitions and metrics play a central role in managing risks and ensuring adequate resilience. Such metrics would provide a sound basis for the development of effective decision-making tools for multihazard environments.

Resilient systems should be developed to meet sustainability requirements defined by the three pillars of sustainability, namely, by reconciling environment, social equity, and economic demands. These three pillars of sustainability are not mutually exclusive but can be mutually reinforcing and often conflicting. Similar to the long-lived and healthy wetlands and forests as sustainable biological systems, humans should sustain their long-term well-being in environmental, economic, and social dimensions and achieve resiliency.

1.2 OBJECTIVE AND SCOPE

This manual of practice (MOP) has a primary objective of providing guidance for the design of hazard-resilient infrastructure systems, and enhancing the resilience of the existing ones for the purpose of meeting community resilience goals. It secondarily contributes to the development and enhancement of standards for infrastructure analysis and design, as well as regulations and building codes that refer to them, in a world in which risk profiles are changing. The MOP and such efforts contribute to achieving hazard-resilient infrastructure.

The MOP is designed to transform our infrastructure and make infrastructure hazard resilient, hence, the title MOP. The authors ruled out using *disaster-resilient infrastructure* as a title because the goal is to manage and act before facing a disaster.

The framework emphasizes infrastructure systems and how they support community resilience. Engineers design long-lived infrastructure.

The footprints of infrastructure have even longer-term influences. The planning and design of new infrastructure should, therefore, account for the projected future conditions related to population growth and changes in hazards including climate of the future. Engineering practice recognizes and accounts for uncertainties in the future conditions; however, the associated uncertainties vary from simple randomness to deep uncertainties associated with projections in the future related to population growth, energy use, and so on.

1.3 INFRASTRUCTURE SYSTEMS AND HAZARDS

For the purposes of this MOP, infrastructure is defined as the basic physical and organizational structures and facilities, for example, building clusters and lifeline systems, needed for operating a society or enterprise. The manual is designed to address all infrastructure types and hazards to provide a basis for developing future manuals of practice that are hazard- or infrastructure-type-specific. The manual covers the technical aspects of infrastructure resilience and relates them to the economics associated with their life cycles, including organizational and socioeconomic considerations.

Ayyub (2014b) provided an asset taxonomy, as shown in Figure 1-1. These assets offer a basis for defining infrastructure systems to meet analytical objectives as set by stakeholders. Figure 1-2 shows a classification of natural hazards. Ayyub (2014b) provided classifications of human-caused hazards that are not a focus of this manual. The concepts covered in this manual are suitable for both natural and human-caused hazards.

Presidential Policy Directive 21 (PPD-21) (PPD 2013): Critical infrastructure security and resilience advance a national policy to strengthen and maintain secure functioning and resilient critical infrastructure. It lists the following 16 critical infrastructure sectors:

1. Chemical sector,
2. Commercial facilities sector,
3. Communications sector,
4. Critical manufacturing sector,
5. Dams sector,
6. Defence industrial base sector,
7. Emergency services sector,
8. Energy sector,
9. Financial services sector,
10. Food and agriculture sector,
11. Government facilities sector,
12. Healthcare and public health sector,
13. Information technology sector,

14. Nuclear reactors, materials, and waste sector,
15. Transportation systems sector, and
16. Water and wastewater systems sector.

These sectors are mappable to the asset taxonomy shown in Figure 1-1. This MOP is intended to be adaptable to cover all these sectors.

Agriculture and Food
- Supply
- Processing
- Production
- Packaging
- Storage
- Distribution
- Transportation

Water
- Dams, wells, reservoirs, and aqueducts
- Transmission pipelines
- Pumping stations
- Sewer systems
- Treatment facilities
- Storage facilities

Public Health
- National strategic stockpile
- National Institutes of Health
- State and local health departments
- Hospitals
- Health clinics
- Mental health facilities
- Nursing homes
- Blood supply facilities
- Laboratories
- Mortuaries
- Pharmaceutical stockpiles

Emergency Services
- Fire houses and rescue
- Federal Emergency Management Agency
- Emergency medical services
- Law enforcement
- Mobile response
- Communications systems

Defense Industry Base
- Supply systems
- Critical R&D facilities

Information and Telecommunications
- Public Switched Telecommunications Network (PSTN)
- Internet
- Switch/router areas
- Access tandems
- Fiber/copper cable
- Cellular, microwave, and satellite systems
- Operations, administration, maintenance, and provisioning systems
- Network operations centers
- Underwater cables
- Cable landing points
- Collocation sites, peering points, and telecom hotels
- Satellite control stations and radio cell towers

Energy

Electricity (Nonnuclear)
- Hydroelectric dams
- Fossil fuel electric power generation plants
- Distribution systems
- Key substations
- Communications

Oil and Natural Gas
- Off shore platforms
- Refineries and pipelines
- Storage facilities
- Gas processing plants
- Product terminals
- Strategic petroleum reserve

Transportation
- Aviation
- Railways
- Highways
- Trucking
- Busing
- Bridges
- Tunnels
- Borders
- Seaports
- Pipelines
- Maritime
- Mass transit

Banking and Finance
- Physical facilities (buildings)
- Operations centers
- Regulatory institutions
- Physical repositories
- Telecommunications networks
- Emergency redundancy service areas

Chemical/Hazardous Materials Industry
- Manufacturing plants
- Transport systems
- Distribution systems
- Storage/stockpile/supply areas
- Emergency response and communications systems

Postal and Shipping
- Processing facilities
- Distribution networks
- Air, truck, rail, and boat transport systems
- Security

National Monuments and Icons
- National parks
- Monuments
- Historic buildings

Nuclear Power Plants
- Commercial owned/operated
- Government owned/operated
- Physical facilities
- Spent fuel storage facilities
- Safety/security systems

Dams
- Large
- Small
- Government owned
- Private/corporate owned

Government Facilities

National Security Special Events

Commercial Assets
- Prominent commercial centers
- Office buildings
- Sports centers/arenas
- Theme parks
- Processing/service centers

Figure 1-1. Asset taxonomy.
Source: Ayyub (2014).

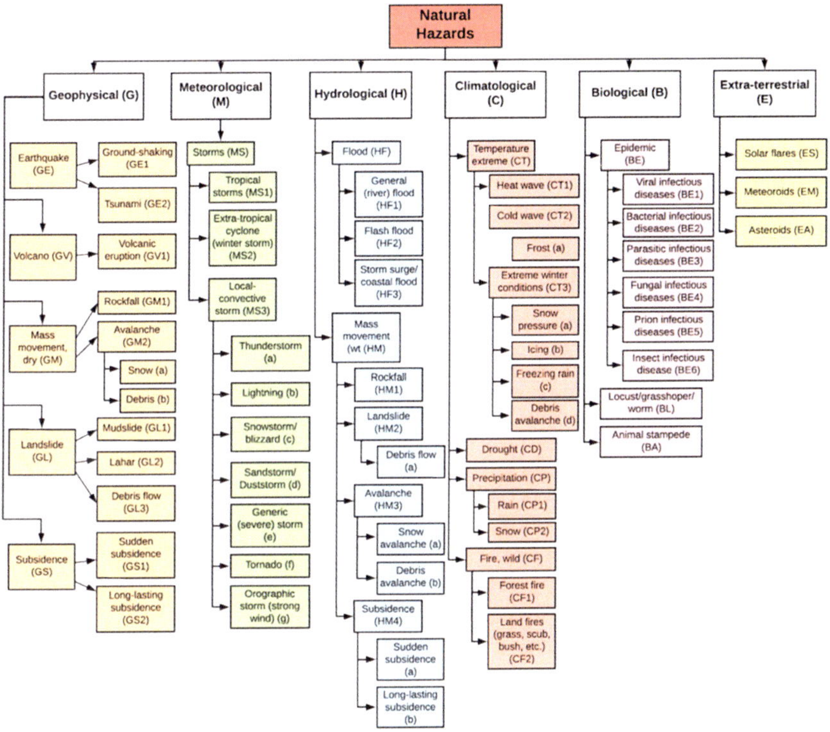

Figure 1.2. Classification of natural hazards.

1.4 STRUCTURE OF THE MOP

Owing to the emerging and evolving nature of topics associated with hazard-resilient infrastructure, this MOP has a varied style that includes step-by-step guidance, research results, and future trends. It covers an emerging field and, therefore, requires the inclusion of both well-established methods and other emerging and promising methods.

The manual consists of the following chapters that are prepared to follow a logical sequence for analyzing and designing hazard-resilient infrastructure:

Chapter 1 is the introduction, which sets the context, articulates the needs, provides an objective and the scope statements for the manual, lists topics warranting additional analysis, and defines users and the uses of the manual. It also defines the relationship between resilience and sustainability and how to consider resilience in the context of sustainability.

Chapter 2 provides a methodology for assessing hazard-resilient infrastructure. This chapter outlines the overall approach and describes its components that are provided in detail in subsequent chapters. It offers a high-level road map for using the manual. It includes several illustrative examples. It is intended to familiarize MOP users with the steps to follow.

The focus of Chapter 3 is on resilience assessment methods of individual facilities and system, system of systems, and networks. It also provides methods for assessing and quantifying performance and resilience of projects and systems including networks. It offers a set of tools to help varied needs. It includes several illustrative examples.

Chapter 4 is on the economics of resilience and risk management. It focuses on methods for economic analysis by accounting for planning horizon and discount rates of resilience related design improvements and available alternatives or technologies.

Chapter 5 builds on previous chapters and offers guidance on designing for resilience, that is, resilience engineering. This chapter suggests approaches to engineer for resilience by bringing concepts from Chapters 3 and 4 together within a planning and design framework. It includes several illustrative examples.

The focus of Chapter 6 is on community socioeconomics. It introduces concepts related to socioeconomics and associated metrics. It focuses on the assessing the socioeconomic role of infrastructure for community resilience.

Chapter 7 surveys emerging resilience-enabling technologies with many examples of current and future technologies that are suitable for designing or retrofitting existing infrastructure.

The manual also includes an appendix on terminology.

1.5 TOPICS WARRANTING ADDITIONAL ANALYSIS

This section lists the topics that warrant additional analysis and development with brief descriptions.

1.5.1 Dependencies and Interdependencies

Infrastructure dependency reflects unidirectional linking of capabilities; for example, Infrastructure A depends on Infrastructure B, where a variation in the latter has the capability to influence some states of Infrastructure A. Dependency can be classified into direct and indirect types. As for the term interdependency, it represents a bidirectional relationship between two or more infrastructures where the state of each infrastructure is influenced or correlated to the state of the other. Hence, Infrastructures A and B are interdependent, meaning A depends on B and, at the same time, B depends on A. Scaling could reveal or mask interdependencies. From a macroperspective, all systems are interdependent; however, this is not necessarily the case in smaller scales where one system may be completely dependent on another. For example, a local water system may depend completely on power for supply, but power may come from faraway and that power being received may have no dependence on the local water system. These dependencies or interdependencies can be a result of physical, cyber, geographical, logical, and social attributes.

1.5.2 Nonstationary Hazards and Adaptive Design Concepts

In 2018, the ASCE Committee on Adaptation to a Changing Climate prepared MOP 140 to address nonstationarity of hazards and adaptive design and risk management (Ayyub et al. 2018). Civil infrastructure systems have traditionally been designed for appropriate functionality, durability, and safety for climate and weather extremes during their full-service lives; however, climate scientists inform us that the extremes of climate and weather have altered from historical values in ways difficult to predict or project. ASCE MOP 140 provides guidance for and contributes to the developing or enhancing of methods for infrastructure analysis and design in a world in which risk profiles are changing and can be projected with varying degrees of uncertainty, requiring a new design philosophy to meet this challenge. The underlying approaches in this manual are based on probabilistic methods for quantitative risk analysis, and the design framework provided focuses on identifying and analyzing low-regret, adaptive strategies to make a project more resilient. It is intended to meet the needs of engineers, researchers, planners, and other stakeholders charged with adaptive design decisions to achieve infrastructure resilience targets while minimizing life-cycle costs in a changing climate. ASCE MOP 140 and this MOP are consistent with each other.

1.5.3 Infrastructure Resilience and Sustainability

The relationship between resilience and sustainability is an ongoing research and attracts professional interest (Ayyub and Wright 2016, Ayyub 2020). This section provides some perspectives for further pursuits.

Resilience and sustainability as system characteristics are necessary for societal endurance and survival. Enhancing them at the element, network, community, and so on levels could lead to massive savings and conservation of resources, not only through efficiencies but also through risk reduction and expeditious recovery in the case of disasters. The rational management of such reduction and recovery is facilitated by practical and fundamental resilience and sustainability metrics. Starting with contrasting resilience and sustainability for understanding their relationships is essential for making appropriate decisions for their control in a cost-effective manner. This chapter introduces concepts that are essential for understanding *sustainable resilience* for hazard-resilient infrastructure systems.

As discussed in Chapter 2, the concept of resilience appears in different domains, ranging from ecology to child psychology and psychiatry to infrastructure systems. It was formally introduced in ecology, defined as the persistence of relationships within a system (Holling 1973), and measured by the ability of the system to absorb change-state and driving variables with persistence. On discussing the philosophical basis of risk analysis, Starr et al. (1976) characterized the resilience of a system as its ability to bounce or spring back into shape or position or to recover its strength or spirit quickly. US federal agencies generally define resilience in accordance with the Presidential Policy Directives (PPD)-8 (PPD 2011) and 21 (PPD 2013). PPD-8 (PPD 2011) defines resilience as "the ability to adapt to changing conditions and withstand and rapidly recover from disruption due to emergencies." PPD-21 (PPD 2013) expands the definition to include "the ability to prepare for and adapt to changing conditions and to withstand and recover rapidly from disruptions. Resilience includes the ability to withstand and recover from deliberate attacks, accidents, or naturally occurring threats or incidents." The Community Resilience Planning Guide for Buildings and Infrastructure Systems National Institute of Standards Technology (NIST 2016) provides a similar definition. Chapter 2 provides additional information on this subject.

Ayyub (2014a, 2015) provided a comparative examination of these definitions and their suitability as a basis for resilience quantification and measurement science. A proposed definition that belongs to the intension class is "the resilience of a system is the persistence of its functions and performances under uncertainty in the face of disturbances" (Ayyub 2014a). For engineering systems, resilience is a system characteristic and the system can be a system of systems. For instance, to be functional after disruption, a building needs communication, power, water, and transportation access for its users as well as has to be functional by itself. Usually, a water utility depends on a power utility to be functional, and a power plant requires cooling water and all depend on natural, human, social, and financial capitals for functionality.

ASCE defines sustainability, in its Policy Statement 418 (ASCE 2016), as a set of economic, environmental, and social conditions in which the whole society has the capacity and opportunity to maintain and improve its quality of life indefinitely, without degrading the quantity, quality, or the availability of natural, economic, and social resources. Sustainable development is the application of these resources to enhance the safety, welfare, and quality of life for the whole society. Several other definitions are available as provided by Webb and Ayyub (2016) based on the definitions provided by EPA (2016a, b, c) and ITA (2016).

Fiksel (2003, 2006) and Redman (2014) provided a discussion on contrasting elements of resilience and sustainability for systems. A resilience theory approach recognizes the following: (1) Change is normal with multiple stable states; (2) experience leads to a gracefully adaptive cycle; (3) resilience originates in ecology for maintaining ecosystem services; (4) results of change are open-ended and emergent; (5) resilience is concerned with maintaining system dynamics; and (6) stakeholder input is focused on desirable dynamics. However, a sustainability science approach recognizes the following: (1) This approach envisions the future and acts to make it happen; (2) it utilizes a transition management approach; (3) it originates in social sciences with the presumption that a society is flawed; (4) its desired results of change are specified in advance; (5) its focus is on interventions that lead to sustainability; and (6) stakeholder input is focused on desirable outcomes. Fiksel (2003, 2006) examined dynamic modeling techniques for examining sustainability, including biocomplexity, system dynamics, and thermodynamic analysis, to investigate the impacts on ecological and human systems of major shifts such as climate change and the associated policy and technology responses. The sustainability science approach of Redman (2014) discusses how an engineer addresses both sustainability and resilience where resilience is treated as an aspect of sustainability during the life cycle of a project or in the management of a system, enterprise, or community. The life cycle includes conception, design, construction, operation, maintenance, and renewal or removal. The necessary steps are as follows: (1) envision the future, for normal function and response to a perturbation, and act for sustainability and resilience; (2) use a transition management approach from current to desired conditions; (3) seek desired results of change that are specified in advance; (4) focus on interventions that lead to sustainability and resilience; and (5) seek stakeholder input by focusing on the desired outcomes. Bocchini et al. (2014) examined the relationships and use of life-cycle costs to unify the treatment of sustainability and resilience in a common risk framework. Gillespie-Marthaler et al. (2019) examined an integrative approach to conceptualize sustainable resilience to enable alignment of adaptation and transformation strategies with desired resilience outcomes. Nelson et al. (2020) presented a high-level, integrated, and dynamic framework for assessing sustainable resilience for complex adaptive systems.

Defining the relationship between resilience and sustainability is context dependent. Two contexts are considered herein: (1) states of infrastructure systems (Ayyub and Wright 2016) and (2) resilience and sustainability requirements for infrastructure systems (Ayyub 2020). The following precepts should be captured in relating resilience and sustainability: (1) systems that are resilient might not be sustainable; (2) systems that are not resilient are not sustainable; and (3) systems that are not sustainable might be resilient.

1.6 USES AND USERS

The MOP can be used to assess existing infrastructure systems, retrofitting methods, and designing new ones. Engineers, researchers, planners, educators, students, and other stakeholders charged with assessment or design decisions would benefit from it to achieve infrastructure resilience targets while minimizing life-cycle costs for a broad range of hazards.

1.7 DATA AND KNOWLEDGE SOURCES

Data and knowledge can be collected and acquired from a wide range of sources based on objective and subjective methods, such as empirical measurements and expert opinions, respectively. Each type entails different uncertainty types and levels. Typically, working data include a mixture as illustrated in Figure 1-3 (Ayyub 2002, 2014b). Analysts and engineers should consider the trade-offs in costs to assess the value of information in informing and affecting decisions and associating benefits and risks.

The National Institute of Standards and Technology (NIST) convened a workshop in 2018 on community resilience data, information, and tools for community resilience planning and decision-making. McAllister et al. (2019) provided a summary of identified gaps and needs.

Emerging applications of artificial intelligence to infrastructure carry the potential for enhancing the quality and reliability of data and information using methods such as neural networks; fuzzy learning from examples and neurofuzzy algorithms to extract features of vibration signals collected from bridges and high-rise building for damage detection and damage pattern recognition as part of structural health monitoring; evolutionary algorithms to solve construction engineering problems; and machine learning to design concrete mixes and predict the behavior of composite structures. These concepts are discussed in Chapter 7.

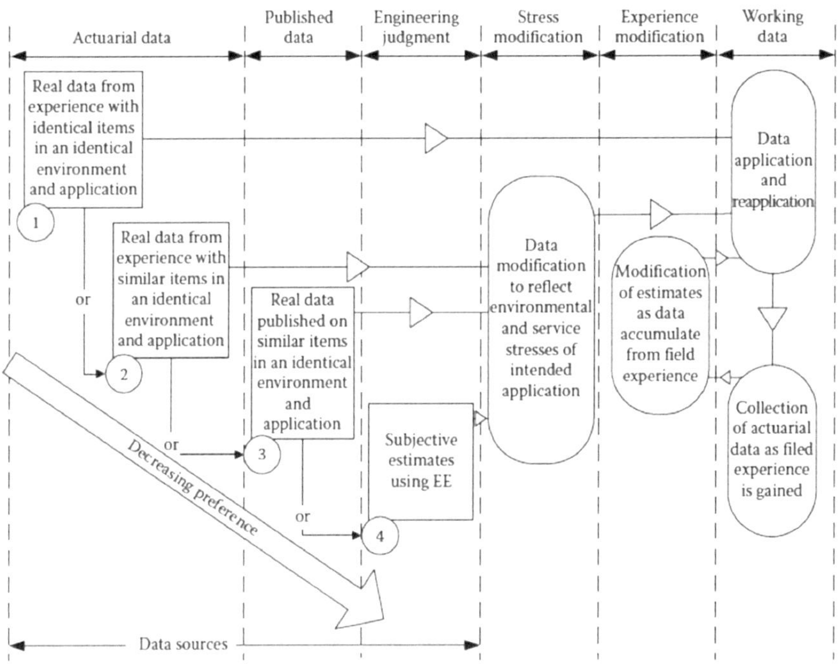

Figure 1.3. Data and knowledge sources and associated reliability from data collection to expert opinion elicitation.
Source: Ayyub (2002, 2014b).

REFERENCES

ASCE. 2009. *Report card for America's infrastructure*. Reston, VA: ASCE.

ASCE. 2016. *Policy statement 418*. Reston, VA: ASCE. Accessed on February 9, 2016. http://www.asce.org/issues-and-advocacy/public-policy/policy-statement-418---the-role-of-the-civil-engineer-in-sustainable-development/.

Ayyub, B. M. 2002. *Elicitation of expert opinions for uncertainty and risks*. Boca Raton, FL: CRC Press.

Ayyub, B. M. 2014a. "Systems resilience for multihazard environments: Definition, metrics, and valuation for decision making." *J. Risk Anal.* 34 (2): 340–355.

Ayyub, B. M. 2014b. *Risk analysis in engineering and economics*. 2nd ed. Boca Raton, FL: Chapman & Hall/CRC Press.

Ayyub, B. M. 2015. "Practical resilience metrics for planning, design, and decision making." *ASCE-ASME J. Risk Uncertainty Eng. Syst. Part A: Civ. Eng.* 1 (3): 04015008.

Ayyub, B. M. 2020. "How are infrastructure resilience and sustainability related?" *ASCE-ASME J. Risk Uncertainty Eng. Syst. Part A: Civ. Eng.* 6(3), Accessed March 8, 2021, https://doi.org/10.1061/AJRUA6.0001067

Ayyub, B. M., H. G. Braileanu, and N. Qureshi. 2012. "Prediction and impact of sea level rise on properties and infrastructure of Washington, DC." *Risk Anal.* 32 (11): 1901–1918.

Ayyub, B. M., M. Medina, T. Vinson, D. Walker, R. N. Wright, A. AghaKouchak, et al. 2018. "Climate-resilient infrastructure: Adaptive design and risk management." In *Committee on adaptation to a changing climate*, edited by B. M. Ayyub. ASCE Manual of Practice 140. Reston, VA: ASCE. https://ascelibrary.org/doi/book/10.10y61/9780784415191.

Ayyub, B. M., and R. N. Wright. 2016. "Adaptive climate risk control of sustainability and resilience for infrastructure systems." *J. Geogr. Nat. Disasters* 6 (2): e118.

Berke, P. R., J. Kartez, and D. Wenger. 1993. "Recovery after disaster: Achieving sustainable development, mitigation, and equity." *Disasters* 17 (2): 93–109.

Bocchini, P., D. M. Frangopol, T. Ummenhofer, and T. Zinke. 2014. "Resilience and sustainability of civil infrastructure: Toward a unified approach." *J. Infrastruct. Syst.* 20 (2): 04014004.

Burby, R., ed. 1998. *Cooperating with nature: Confronting natural hazards and land-use planning for sustainable communities.* Washington, DC: Joseph Henry Press.

EPA (Environmental Protection Agency). 2016a. "Learn about sustainability." Accessed February 9, 2016. http://www.epa.gov/sustainability/learn-about-sustainability.

EPA. 2016b. "EPA's report on the environment." Accessed February 9, 2016. http://cfpub.epa.gov/roe/chapter/sustain/index.cfm.

EPA. 2016c. "Green buildings." Accessed February 9, 2016. http://archive.epa.gov/greenbuilding/web/html/faqs.html.

Fiksel, J. 2003. "Designing resilient, sustainable systems." *Environ. Sci. Technol.* 37 (23): 5330–5339.

Fiksel, J. 2006. "Sustainability and resilience: Toward a systems approach." *Sustainability: Sci. Pract. Policy* 2 (2): 14–21.

Gilbert, S. W. 2010. *Disaster resilience: A guide to the literature.* NIST Special Publication 1117. Gaithersburg, MD: Office of Applied Economics, Engineering Laboratory, National Institute of Standards and Technology.

Gillespie-Marthaler, L., K. S. Nelson, H. Baroud, D. S. Kosson, and M. Abkowitz. 2019. "An integrative approach to conceptualizing sustainable resilience." *Sustainable Resilient Infrastruct.* 4 (2): 66–81.

Holling, C. S. 1973. "Resilience and stability of ecological systems." *Annu. Rev. Ecol. Syst.* 4: 1–23.

ITA (International Trade Administration), Department of Commerce. 2016. "How does commerce define sustainable manufacturing?"

Accessed February 9, 2016. http://www.trade.gov/competitiveness/sustainablemanufacturing/how_doc_defines_SM.asp.

McAllister, T., C. Clavin, B. Ellingwood, J. van de Lindt, D. Mizzen, and F. Lavelle. 2019. *Data, information, and tools needed for community resilience planning and decision-making*. NIST Special Publication 1240. Gaithersburg, MD: National Institute of Standards and Technology.

Nelson, K., L. Gillespie-Marthaler, H. Baroud, M. Abkowitz, and D. Kosson. 2020. "An integrated and dynamic framework for assessing sustainable resilience in complex adaptive systems." *Sustainable Resilient Infrastruct.* 5 (5): 311–329.

NIST (National Institute of Standards and Technology). 2016. *Community resilience planning guide for buildings and infrastructure systems*, Vol. 1. NIST Special Publication 1190. Gaithersburg, MD: NIST.

NRC (National Research Council). 2012. *Disaster resilience: A national imperative*. Washington, DC: National Academies Press.

PPD (Presidential Policy Directives). 2011. "National preparedness." PPD-8. Accessed February 9, 2016. http://www.dhs.gov/presidential-policy-directive-8-national-preparedness.

PPD. 2013. "Critical infrastructure security and resilience." PPD-21. Accessed June 25, 2013. http://www.whitehouse.gov/thepress-office/2013/02/12/presidential-policy-directive-criticalinfrastructure-security-and-resil.

Redman, C. L. 2014. "Should sustainability and resilience be combined or remain distinct pursuits?" *Ecol. Soc.* 19 (2): 37.

Starr, C., R. Rudman, and C. Whipple. 1976. "Philosophical basis for risk analysis." *Annu. Rev. Energy* 1: 629–662.

UNDRR (United Nations Office for Disaster Risk Reduction). 2012. *Making cities resilient: My city is getting ready! A global snapshot of how local governments reduce disaster risk*. Geneva: United Nations. Accessed January 31, 2021. https://www.unisdr.org/campaign/resilientcities/home/article/moving-to-the-decade-of-action-with-mcr2030.

UNDRR. 2015. *Sendai framework for disaster risk reduction 2015–2030*. 1st ed. Geneva: United Nations. Accessed April 7, 2019. https://www.unisdr.org/files/43291_sendaiframeworkfordrren.pdf.

United Nations. 2016. "The infrastructure—inequality—resilience Nexus." Chap. 2 in *The global sustainability development report*. New York: Dept. of Economic and Social Affairs.

Webb, D., and B. M. Ayyub. 2016. "Sustainable construction and manufacturing. 1: Definitions, metrics, and valuations for decision making." *ASCE-ASME J. Risk Uncertainty Eng. Syst. Part A: Civ. Eng.*

CHAPTER 2

A METHODOLOGY FOR ASSESSING HAZARD-RESILIENCE INFRASTRUCTURE

2.1 INTRODUCTION

The concepts of resilience in the engineering community can be considered to build on risk principles. However, the concept tends to deal less with factors extrinsic to the system, rather than the intrinsic properties of the system. At a high level, and without credence to nearly two decades of research on the topic, resilience reflects the ability of a system to withstand and recover from some precipitating event. The concept of resilience was initially introduced by Holling (1973) to describe the ability of an ecosystem to persist over time despite external stressors. The ecosystem may evolve or adapt to the stressor, and the goal was not to return to the original state but rather to a state better suited for the conditions. Because the concept of resilience has been adopted by other fields, including engineering and childhood psychology, the focus of this discussion is limited to infrastructure resilience to extreme events. Broadly, resilience has been defined as follows:

- "The ability to prepare and plan for, absorb, recover from, or more successfully adapt to actual or potential adverse events" (NRC, 2012).
- "The ability to prepare for and adapt to changing conditions and withstand and recover rapidly from disruptions. Resilience includes the ability to withstand and recover from deliberate attacks, accidents, or naturally occurring threats or incidents" (PPD-21) (PPD 2013).
- "The ability to prepare for and adapt to changing conditions and withstand and recover rapidly from disruptions" (Ayyub 2014a, b).

This definition continues to identify what could threaten a system. They include deliberate attacks, accidents, and naturally occurring hazards.

- "The capability to mitigate against significant all-hazards risks and incidents, and to expeditiously recover and reconstitute critical services with minimum damage to public safety and health, the economy, and national security" (Policy Statement 518) (ASCE 2013).
- "The capacity to withstand or absorb the impact of a hazard through resistance or adaptation, which enable it to maintain certain basic functions and structures during a crisis, and bounce back or recover from an event" (UNISDR, 2012).
- Davis et al. (2018) defined a resilient infrastructure system as one that is "managed to provide safe and reliable services to customers, cope with chronic stressors, and accommodate hazard-related impacts with ability to continue providing services or limit service outage times tolerable for community recovery efforts." This definition for infrastructure resilience is consistent with the resilience definitions referenced previously but has a strong emphasis on the services provided by infrastructure systems and helps infrastructure organization's focus on operationalizing resilience into practice (Davis et al. 2018). However, the definitions referenced previously have a high-level focus, establishing a broad-level policy to establish a consistent direction for infrastructure systems, social institutions, and others, and are not always easy to understand how to implement directly into practice. As an alternative, Davis et al. (2018) offered a basis for a definition as follows: A resilient infrastructure or lifeline system has the ability to accommodate hazard-related impacts and continue providing services or limit service outage times tolerable for community recovery efforts.

There is significant convergence in these definitions. Resilience in the engineering and infrastructure field is generally viewed as the ability of a system to withstand the impact of some extrinsic event and to recover from it. The intension of the word *resilience* includes intrinsic properties, including robustness, redundancy, resourcefulness, adaptability, and rapidity (Bruneau et al. 2003). Each influences a system's ability to be resilient. They are defined as follows:

- **Redundancy**: The ability of a system to function even after some of its components or subsystems have failed, or more broadly, the extent to which the system satisfies and sustains functional requirements after the failure of some of its components or subsystems (Biringer et al. 2013).

- **Robustness**: The ability of elements or systems to withstand a significant level of performance demand without suffering a loss of function.
- **Rapidity**: The capacity of a system to reduce losses and meet performance requirements by recovering in a timely manner (Biringer et al. 2013).
- **Resourcefulness**: The capacity of a system to mobilize materials and/or human resources to recover and meet performance goals after disruption event (Biringer et al. 2013).

The definition of resilience adopted for this work comes from PPD-21: "the ability to prepare for and adapt to changing conditions and withstand and recover rapidly from disruptions. Resilience includes the ability to withstand and recover from deliberate attacks, accidents, or naturally occurring threats or incidents" (PPD 2013). According to Ayyub (2014b), "the persistence of [a system's] functions and performances under uncertainty in the face of disturbances" is beneficial as it harkens back the original definition of resilience relating to the persistence of ecological systems and does not provide strict requirements for a system to return to its original state before the disaster but rather possibly to a better or evolved state. This definition is relevant to a wide array of systems affected by hazards, including both infrastructure and communities. The three keywords in this definition are performance, uncertainty, and persistence. Performance is in reference to some objective measure of system function, such as structural integrity, life cycle costs, or even product quality. Uncertainty relates to the hazards that may transpire or the scenarios that may emerge. This could include banal episodes or black swans. Persistence is in the relation to how the system continues or recovers in the face of the hazard. The one key difference in this definition compared with other definitions is the exclusion of the word *ability*. By excluding this word, it encourages the development of metrics that are less process focused and that are more outcome focused.

This chapter provides the methodology structure for assessing the hazard resilience of infrastructure. Subsequent chapters provide details and examples. Appendix A provides definitions of additional terminology.

2.2 INFRASTRUCTURE AND LIFELINE SYSTEMS

The scope of infrastructure resilience is very broad and covers the systems providing services within large and small urban areas as well as rural areas. It also covers those systems providing services in support of other supply chain systems of goods and services, which may or may not originate within the communities. This includes supply chains such as agriculture and food production.

2.3 OVERALL METHODOLOGY

This section provides a high-level methodology for resilience assessment, a basis for analysis and design, as shown in Figure 2-1. It consists of the following primary steps as provided in the primary boxes in the figure listed herein in an order that reflects a suggested process flow for assessing the resilience of an infrastructure system:

1. Context definition,
2. Hazard identification and characterization,
3. Failure probability estimation and fragility curves,
4. Resilience assessment,
5. Exposure and loss analysis,
6. Economic valuation and loss accumulation,
7. Risk quantification as loss exceedance rates or probabilities,
8. Extremes and uncertainty analysis,
9. Resilience engineering and design,
10. Life cycle analysis,
11. Risk-informed decision analysis for resilience engineering, and
12. Community socioeconomics.

These steps are color coded in the figure according to the legend provided in the figure to reflect several key themes consisting of (1) selecting or setting an initial design; (2) computing probabilities as needed; (3) estimating potential failure consequences; (4) tracking and providing information flow; and (5) accounting for other considerations, such as planning horizon, discount rate, other social and economic attributes, and so on. The methodology may also include other considerations that are not shown in the figure, such as information and data sources and governance. The methodology is a top-down approach that starts with the hazard scenarios and predictions or projections. The steps are described briefly in subsequent sections.

Figure 2-1 provides the key steps necessary for assessing infrastructure systems. These steps define a workflow process and can be used to identify appropriate tools and outcomes for system resilience assessment and management. The outcomes are useful for analyzing and designing systems or addressing recovery and retrofit needs after actual disruptions.

For this methodology, an infrastructure system is defined as the basic physical and organizational structures and facilities (e.g., building clusters and lifeline systems) and their associated organizational entities needed to operate a society or enterprise. The methodology emphasizes infrastructure systems and how they support community resilience.

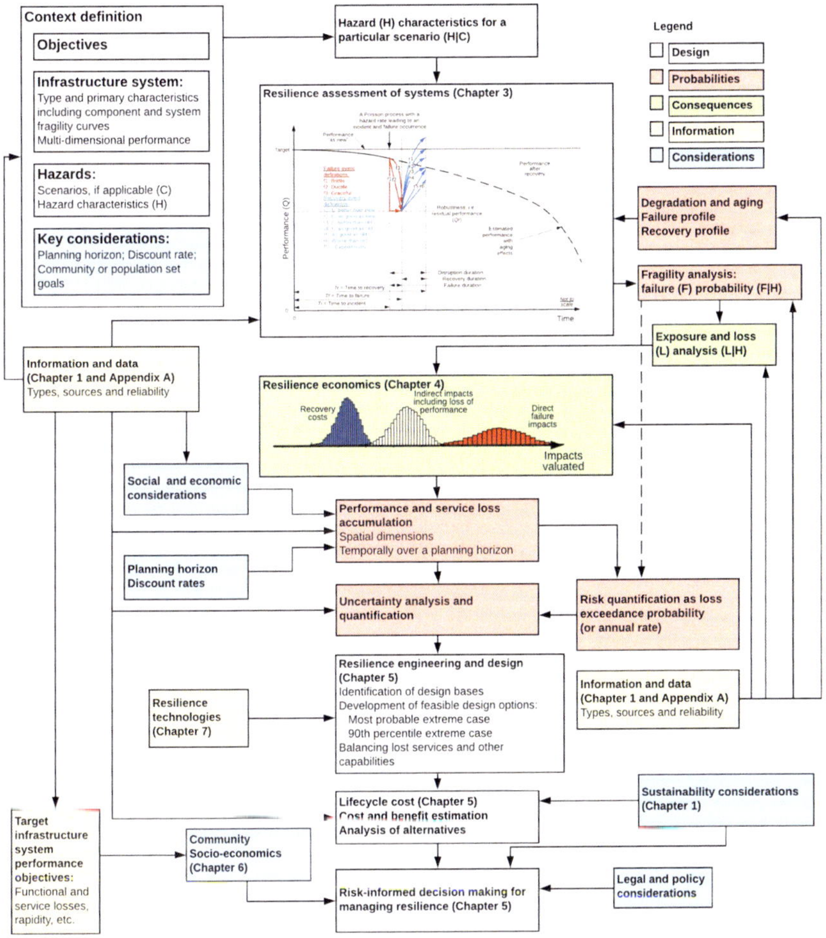

Figure 2-1. Methodology for resilience assessment for analysis and design.

2.3.1 Context Definition

The context should be defined for a particular system under consideration based on its planning time horizon and performance requirements. A suitable hazard-specific context could be identified to include selecting appropriate future projections related to hazards and needs, such as related changes in population, economy, technology, energy, land use, agriculture, and so on. The uncertainty associated with this context definition requires analysis and quantification. More than one context can be identified and used, if needed.

A project-specific context includes selecting project-related key parameters, such as system requirements and attributes, hazard characteristics, and an appropriate planning time horizon or design life, discount rate, and so on.

The methodology applies to all infrastructure systems but is written in a manner that addresses one system at a time. For all infrastructure systems to uniformly support the overall community resilience, the performance targets for each system should be incorporated in a system-of-systems approach, as discussed in Chapter 3.

Resilience is a multifaceted notion and the aggregate of many dimensions. The ASCE Infrastructure Resilience Division (IRD) has defined an infrastructure resilience domain as shown in Figure 2-2 (Davis et al. 2018). The domain is structured around three primary dimensions: civil infrastructure and lifeline systems (shown on the vertical axis), hazards (shown on the horizontal axis to the right), and the adverse event cycle (shown on the horizontal axis pointing out of the plane of the paper). These primary dimensions make up a matrix of resilience cubes as schematics, with each cube having a system–hazard–event cycle locus. Figure 2-2 explicitly outlines the cube for communication system-flood-preparation. Each cube identifies the dimensions needed to be addressed to achieve true infrastructure resilience; for each system, all hazards of importance and every step in the event cycle need to be addressed.

Figure 2-2 identifies the infrastructure systems on the vertical axis, which are broken down into two primary categories: lifeline systems and building clusters. The traditional lifelines are itemized on the vertical axis as communication; electric power; water; wastewater; inundation protection, for example, covering flooding, debris flow, avalanche, and so on; natural gas; liquid fuels; solid waste management; and transportation systems. Other lifeline systems may need to be included based on the local community needs. For example, in New York City, Consolidated Edison operates the largest steam system in the world, providing services to 1,700 customers for commercial and residential establishments, serving more than 3 million people, used for heating, cooling, cleaning, and disinfection (conEdison 2018, Brown 2018); the steam system is considered a local lifeline system. Building clusters are defined by NIST (2015) as "a set of buildings—and supporting infrastructure systems—that serve a common function such as housing, healthcare, retail, etc. Clusters are not necessarily geographically co-located, and may be distributed throughout the community." All engineering disciplines are needed in support of these systems.

Infrastructure systems are large and complicated geographically distributed networks, as discussed in Chapter 3, made of numerous interlinked components that are built over long periods of time, under a wide variety of conditions, using many different materials and construction techniques. The systems include all subsystems, regardless of who owns or operates them, from source to delivery, from collection to disposal, and so on. Some parts of each system are more fragile than others. The vulnerabilities of each system are dependent on the geospatial layout of the system and the distribution of vulnerable components relative to the geographic distribution of the hazards the systems are exposed to. These large systems traverse many geologic formations and hazards (e.g., rock, soil, landslides, rivers, and reclaimed land). Severe hazards can affect large areas with varying intensity resulting in a large variation of effects (e.g., flooding or landslides in some locations but not in others).

Within each cube, there are other dimensions and characteristics of resilience. This is represented in Figure 2-2 by dividing each cube into smaller volumes. For example, redundancy, resourcefulness, rapidity, and robustness (Bruneau et al. 2003) are recognized as important resilience properties as well as adaptability and the technical, organizational, social, economic resilience dimensions (Bruneau et al. 2003) that are essential for understanding infrastructure resilience. These properties and domains are embedded within each of the cubes making up Figure 2-2.

The figure also shows the geopolitical hierarchy in which issues may reside at each locus. For each infrastructure type across the nation, there are different hazards of importance depending on the geographic location. The infrastructure systems themselves span from the local, regional, state, national, to the international extent, and the issues to be managed fall within the local, regional, state, federal, and/or international levels. The hierarchy of issues to deal with could be local to international regardless of the system size and the customers they provide services to. The geopolitical dimension is represented as a smaller volume within the cube shown in Figure 2-2.

Some resilience dimensions may be located only within some cubes shown within Figure 2-2 but not in others. For example, the project life cycle primarily falls within the mitigation and rebuild steps of the event cycle, whereas the other steps in the event cycle are mainly intended to be continuous preparation/implementation programs. Although not fully sketched out in Figure 2-2, the project life-cycle dimension is represented as a smaller volume within each respective cube.

There are numerous crosscutting and external aspects to infrastructure resilience. Several of the crosscutting items are shown around the perimeter of the 3D model (to the right and top) and include all-hazards, multihazards, dependency relationships, postdisaster investigations, data collection, policy, regulation, socioeconomics, research, and development. These topics are shown in Figure 2-2 along the axes most applicable. Risk and uncertainty permeate throughout Figure 2-2. Uncertainty is involved with every hazard and step within the event cycle important to understand to make effective use of the matrix and improve infrastructure system resilience.

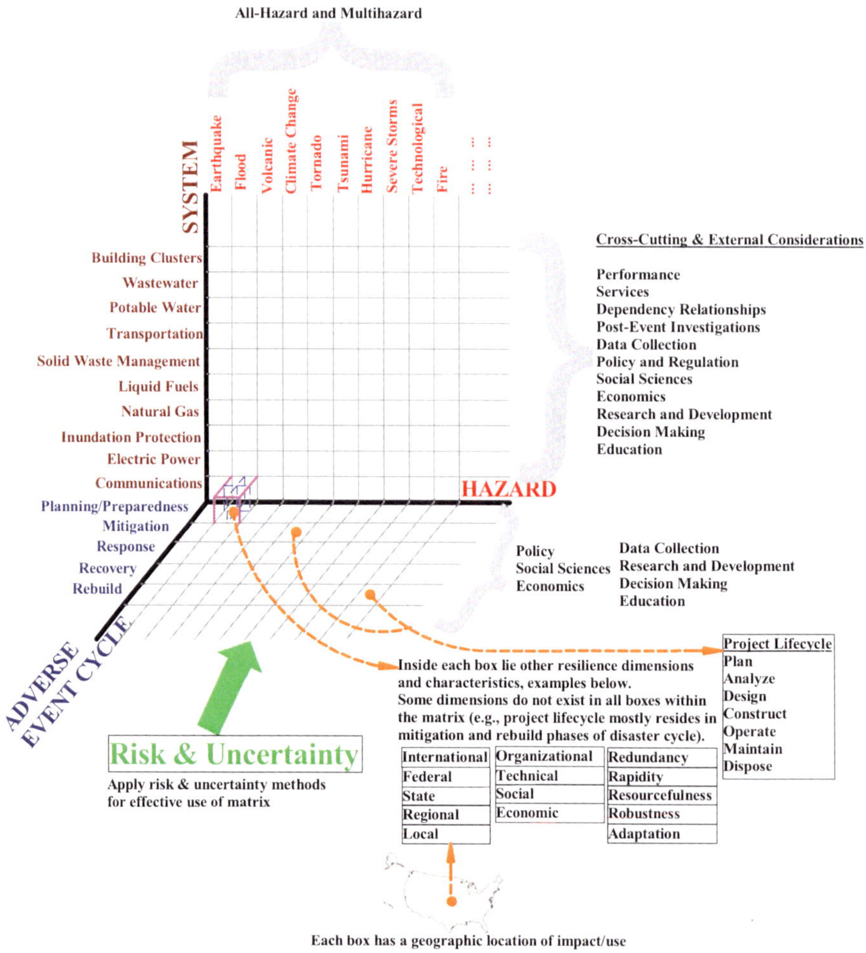

Figure 2-2. ASCE infrastructure resilience domain.
Source: Modified from Davis et al. (2018).

Decision- and objective-related contexts include decision criteria, the risk attitude of a decision maker, and any budgetary- and policy-related constraints. Davis et al. (2018) provided an example of how the resilience model can be reframed in terms of the resilience domains and properties to identify general characteristics of resilient lifeline systems (Figure 2-2).

2.3.2 Hazard Identification and Characterization

The list of hazards requiring consideration is context dependent. Chapter 1 describes key hazards. Typically, the selection of applicable hazards is based on infrastructure and lifeline system type, geographic location, planning time horizon, project requirements, and any budgetary- and policy-related constraints. Quantifying probabilities, rates, and intensities of hazards requires the use of uncertainty analysis, stationary or nonstationary stochastic processes, prediction methods, and extreme value analysis. Some of these methods are described in subsequent chapters. Uncertainty quantification is necessary to establish bounds on estimates, such as confidence intervals or interval probabilities. Some of the information needed might be unavailable. In such cases, expert opinion elicitation could be employed (Ayyub 2001).

2.3.3 Failure Probability Estimation and Fragility Curves

Estimating failure probabilities associated with hazards or an event of interest or a scenario consisting of a series of events with associated dependencies, required for a particular decision situation, depends on the information available. Methods can be classified as analytical approaches, statistical techniques, and Bayesian treatments that bring together subjective and objective information. Methods for computing failure probabilities and developing fragility curves are well established, see, for example, Ayyub (2014a) and Shinozuka et al. (2000). A fragility curve for a system shows the probability of exceeding a particular damage state (or performance level including failure) as a function of a demand parameter on the system that represents the hazard or event intensity level, such as ground motion from earthquakes, for example, spectral displacement at a given frequency.

Figure 2-3 shows the performance of a system over time, as provided by Ayyub (2015). System performance is commonly equated to the quality of infrastructure or functionality by many researchers (Bruneau et al. 2003, Ayyub 2014a, Cimellaro et al. 2010, Attoh-Okine et al. 2009, Tierney and Bruneau 2007). However, performance of a system is the aggregate of the entire set of component performances and how the organization interacts with and manages them at the component and system levels. As shown in Figure 2-3, performance gradually changes over time, for better or worse in the case of aging as an example, and sometimes abruptly. Abrupt change results in some level of performance failure associated with a significant incident or disastrous event such as a hazard strike. The event intensity may be of any magnitude.

2.3.4 Resilience Assessment

Figure 2-3 represents the performance of any system and provides analytical and computational frameworks (Ayyub 2014b). An abrupt loss of performance is the result of a hazard strike. The time to recovery following

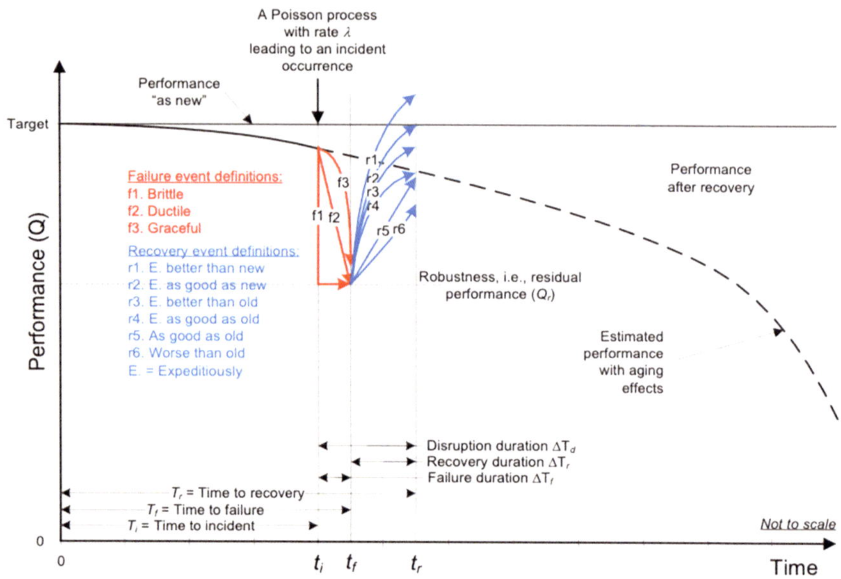

Figure 2-3. Infrastructure (building or lifeline) system performance over time. Source: Ayyub (2014b).

an abrupt performance loss is a result of planning, preparedness, response, recovery, and rebuild capabilities. The system failure may be brittle, ductile, or graceful. Recovery may be worse than old, as good as old, expeditiously as good as old, better than old, as good as new, or better than new. In extreme cases, recovery may include complete abandonment and rebuild, as was the case for all systems in some towns in China following the 2008 Wenchuan earthquake and in Japan following the Great East Japan Earthquake and Tsunami. The curves in Figure 2-3 involve the technical and organizational dimensions of systems requiring the physical systems and the organizations that manage them to withstand hazard-related impacts and recover quickly from them. Sociotechnical systems emerge from the interplay of the technical and organizational dimensions combined with how they interact with social dimensions of resilience (Davis et al. 2018). Van der Merwe et al. (2018) presented a framework for assessing the resilience of sociotechnical systems. The technical and organizational aspects are equally important for creating resilient infrastructure systems.

The performance of each infrastructure system is dependent on the performance of other systems. Impacts resulting in loss of performance may result from direct system damages as a result of a hazard strike or from loss of dependent services even if the system was not damaged (e.g., loss of electric power may impact water system performance because of the inability to pump).

This assessment should also consider the potential for secondary damage from hazards that may result as a cascading effect of the primary damage inflicted by an initial hazard strike. For example, if an earthquake was to cause a dam to break, then the massive flood wave from the reservoir causes severe secondary damage to the infrastructure in the inundation zone; in case of a radiation release from a nuclear power plant, the radiation spread would contaminate the regional environment.

Models for resilience assessment are covered in detail in Chapter 3.

2.3.5 Exposure and Loss Analysis

Failures result in consequences and associated severities. Their extent depends on the exposure of populations, property, and the environment as a result of losses from system service provision and operability. Service losses result from damages to the system from a hazard strike and/or from loss of service from other dependent systems, for example, gas-fired power-generation plant having loss of fuel service. Losses herein are directly related to the system; and in the next section, larger community-wide social and economic losses resulting from the infrastructure system service losses are addressed.

Every infrastructure system provides services to the community defined by its functionality and operability, where functionality is the quality or state of working properly to provide a regular reliable service at, or as close as possible to, what the infrastructure system provided prior to an event, and operability is the fitness, capacity, or ability to use to provide basic services allowing customers/users to receive normal, or near normal, amenities from a potentially impaired infrastructure system following an event (Davis 2019).

For hospital and healthcare building clusters, the services are related to health care; for water systems, the services are related to water usage, and similarly for other infrastructure systems. The services provided by each system can be defined in terms of the basic service categories essential for supporting community resilience. The ability to continue providing these service categories, or rapidly restore them if they are lost after a hazard strike, is related to system performance or functionality, which is dependent on the technical and organizational dimensions. Postevent performance or functionality is measured relative to pre-event normal functionality, which is assumed to be 100%. Loss of functionality following an event results in a measure of less than 100% until it is completely restored to pre-event levels. In this context, as an example, loss of power to a hospital is considered a reduction in functionality below 100%, even if the power is temporarily supplied through emergency generation; the hospital building system is not considered 100% functional until the electric power is restored. Improved functionality following restoration and mitigation after an event can result in a functionality measure of greater than 100%. A highly resilient system can be operable and achieve the provision of the core services to the community prior to achieving full functionality restoration. The system is considered to have 100% operability when all the core community services are met, which for a highly resilient system will occur prior to returning to 100% functionality. Using the same example for power loss to a hospital, the building system is considered 100% operable when using the generator to provide power if it can still achieve the provision of all core community services prior to the electric power service being restored; that is, if the main power supply goes down as the result of an event, functionality is diminished, but operability can be maintained by a backup generator.

Figure 2-4 presents a more complicated example of the Los Angeles Water System following the 1994 Northridge Earthquake. System performance, in this case, is multidimensional. The water system service categories, treated as performance dimensions per Figure 2-3, are defined as follows (Davis 2014):

- **Water Delivery**: This service is fully achieved when the system is able to distribute water to customers, but the water delivered may not meet quality standards (requires public notification), pre-event volumes (requires water rationing), or fire flow requirements (affecting firefighting capabilities).
- **Water Quality**: This service is fully achieved when the water quality at customer connections meets pre-event standards. Potable water meets the health standards (public notices for water use are removed), including minimum pressure requirements to ensure contaminants do not leach into the system.
- **Water Quantity**: This service is fully achieved when the water flow to customers meets the pre-event volumes (water rationing removed).
- **Fire Protection**: This service is fully achieved when the system is able to provide pressure and flow of a suitable magnitude and duration to fight fires.

Figure 2-4 plots the service category loss and restoration along with functionality. Functionality is calculated using Davis (2013). Operability is plotted as the accumulation of all the service categories. As seen in Figure 2-4, operability was returned in 12 days after the earthquake, but the functionality was not restored for another 6 years (Davis 2014). At 12 days after the event, all customers were able to resume the normal use of water services, even though the water system was not functioning as it was prior to the event. This is a result of organizational and technical resilience characteristics, primarily having a well-trained and equipped staff possessing the ability to rapidly mobilize to the hazard strike and adapt to the situation and a highly redundant system with isolation capabilities.

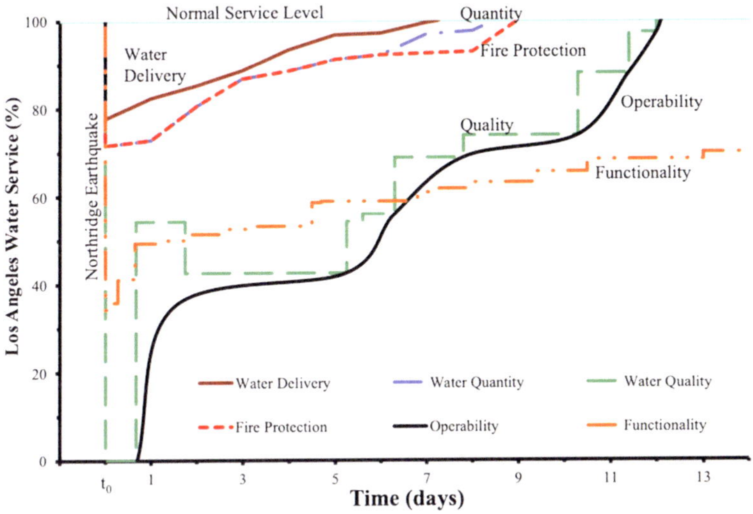

Figure 2-4. Los Angeles water system service restorations following the 1994 Northridge Earthquake. Operability is not a service; it is a milestone delineating collective restoration of water delivery, quality, quantity, and fire protection services.
Source: Davis (2014).

From this example, it is clear how operability is a measure of how the infrastructure system supports community resilience and functionality is a measure of the infrastructure system resilience.

2.3.6 Economic Valuation and Loss Accumulation

Economic valuation is necessary to accumulate the losses in monetary terms over a planning horizon using an appropriate discount rate. Ellingwood and Lee (2016) provided discussion that could guide analysis for intergenerational life-cycle risk assessment of civil infrastructure exposed to hurricanes under climate change.

Accumulating the loss estimates over all the scenarios with an appropriate discount rate (i) over a planning horizon (T) produces a total loss random variable (Gilbert and Ayyub 2016).

The social and economic activities of a community should be examined. As related to infrastructure resilience, it includes the social and economic aspects supported directly and indirectly by the infrastructure systems. As previously defined, infrastructure systems are needed for the operation of society and enterprise. Without them, there is a loss of social and economic functions. The increased ability to maintain services and rapidly restore any lost services, and provision of continuity during service losses, improve the resilience of social and economic systems. In addition, the planning, preparedness, mitigation, emergency response, recovery, and rebuild capabilities of social and economic institutions (including government and businesses), along with financial mechanisms such as insurance and loans, are essential to assessing community resilience. Understanding these interactions is vital to establish community performance targets.

2.3.7 Risk Quantification as Loss Exceedance Rates or Probabilities

Understanding risk is key for addressing the potential of hazards impacting a system. Risk can broadly be defined as a potential for a consequence or the effect of uncertainty on objectives (Ayyub 2014a). Natural hazards have the potential for inducing losses. In the context of natural hazards, the risk to a system (where the system is a community, infrastructure, cybersystem, or something else) is ostensibly extrinsic; the hazard's forces threaten the system. However, a deeper investigation reveals that risk to systems from natural disasters is similarly intrinsic. There are factors from within a system that can augment risk from external factors. These include permissive land-use practices, poor maintenance practices, and rapid urbanization. Risk is commonly described by a triplet (Kaplan and Garrick 1981): What can happen? How likely is it that will happen? If it does happen, what are the consequences? Understanding system vulnerabilities that reflect a system's intrinsic lack of ability to withstand hazards is necessary for risk analysis.

Quantifying risk for a system brings together the probabilities and consequences in terms of an accumulated loss (L) random variable as follows:

$$L = \sum_{t=1}^{T} \left(\sum P(E) P(H \mid E) P(F \mid H)(L \mid F) e^{-it} \right) \tag{2-1}$$

where

L = Loss (L) accumulated over the planning horizon represented by the time period T,
$P(E)$ = Probability of an event (E) or related scenario at time t,
$P(H \mid E)$ = Annual probability of a hazard (H) under the conditions defined by E,
$P(F \mid H)$ = Probability of a failure (F) on the occurrence of H,
$L \mid F$ = Loss (L) on the occurrence of F, and
I = Annual discount rate.

Equation (2-1) is based on several key simplifying assumptions that include (1) the occurrence of a single event (E) impacting a system in a year associated with a particular intensity of a hazard (H) type such as an earthquake, flood, and tornado; (2) discretization of the planning horizon (T) by years; (3) recovery of the system after E, that is, not to consider another event impacting the system prior to its recovery to any prescribed states, as illustrated in Figure 2-3; and (4) not accounting for changes in things exposed and impacted by the failure of the system, such as surrounding asset inventory impacted by the failure.

Accumulating the loss according to Equation (2-16) over all the scenarios, that is, as represented by the inner summation sign, with an appropriate discount rate (i) over a planning horizon (T), that is, as represented by the outer summation sign, produces a total loss random variable (L) (Gilbert and Ayyub 2016).

2.3.8 Extremes and Uncertainty Analysis

The examination of disasters and designing disaster-resilient infrastructure require the analysis and characterization of extremes in probabilistic terms. The analysis of extreme should be based on an examination of uncertainties associated with hazards.

Uncertainty is the state of deficiency in information with two uncertainty types identified as follows: (1) Aleatory uncertainty is the inherent, random, or nonreducible uncertainty, such as material strength randomness. (2) Epistemic uncertainty is the knowledge-based uncertainty that can be reduced with the collection of data or attainment of additional knowledge (Der Kiureghiana and Ditlevsen 2009, Ayyub 2014a).

The uncertainty associated with some events might not be quantifiable completely, and, therefore, accounting for it in engineering practice would require an appropriate understanding and treatment of uncertainty including engineering judgment. In general, uncertainty sources can be broadly classified for convenience into the following types (Ayyub and Klir 2006, Ayyub et al. 2018):

1. Uncertainty that is recognized and well characterized, such as material properties;
2. Uncertainty that is recognized but is moderately characterized, such as future precipitations, hurricanes, wind speed, and so on;
3. Uncertainty that is recognized, however, is poorly characterized, such as the future energy use by populations worldwide;
4. Uncertainty that is recognized, however, cannot be characterized, such as global governance and cooperation; and
5. Uncertainty sources of an unknown existence or nature, for example, physical laws or behaviors that are not discovered yet or undiscoverable based on the ongoing intellectual pursuits.

Several approaches are available to address these uncertainty sources, as discussed in the subsequent section.

2.3.9 Resilience Engineering and Design

Engineering (or designing) for resilience can only become a reality if resilience is not only treated as a noble concept but also as a system's characteristic that can be quantified and measured, so that explicit performance targets can be established for the lifetime, or the target design life, of the system. Standard measures of resilience are provided in Chapter 3; however, they are evolving. They tend to be system and hazard dependent. In general, conventional design approaches have involved providing a minimum level of desired reliability, robustness, and redundancy via target strength and serviceability criteria that must be met. During the last few decades, performance-based design approaches such as performance-based earthquake engineering have incorporated strategies for probabilistic assessment of direct economic loss and collapse safety because of earthquakes. In many parts of the country, especially in the US West Coast, the application of performance-based earthquake engineering concepts to life safety and collapse prevention has become more prevalent throughout the years.

Chapter 5 is devoted to designing for resilience, with several illustrative examples for the protection of equipment and buildings exposed to flood and new construction in seismic-prone areas. The treatment of socioeconomic impacts is covered in Chapter 7, where the focus is strictly on resilience engineering. The examples provided in Chapter 5 are by no means comprehensive in terms of covering all components associated with resilience-based design. They are presented to illustrate, albeit in a rather simple manner, the application of some of the fundamental design basis and principles associated with resilience engineering. The intent is to familiarize the reader with fundamental ways in which resilience engineering has been and can be incorporated into the current design practice.

2.3.10 Life-Cycle Analysis

The life-cycle cost of an infrastructure system, such as a bridge, is the present value total costs of planning, designing, constructing, operating, maintaining, rehabilitating, and decommissioning the system. Life-cycle cost analysis (LCCA) is a recognized analytical process that supports infrastructure and lifeline-related decisions. Investments typically involve large initial capital outlays, followed by subsequent smaller costs and disbursements that may represent sustainment, restoration, or modernization investments as illustrated in Figure 2-5 in terms of a cash flow over time using a convenient time step, such as a year. The figure shows the costs, that is, investments, as downward arrows. The upward arrows show the benefits gained on an annual basis. Typically, engineers and analysts focus on the costs, and the benefits are not expressed explicitly and considered to be a

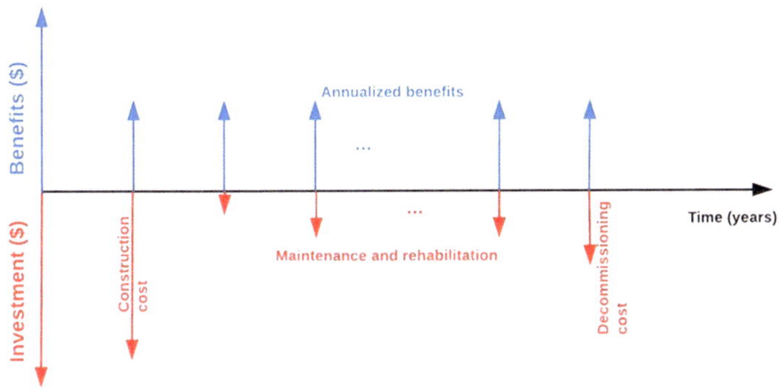

Figure 2-5. Investments and benefits over a system's life cycle.

function of the costs based on economic concepts such as the willingness to pay as a basis for economic valuation of these benefits. Life cycle analysis should be included in the resilience engineering and management, as discussed in Chapter 5. Frangopol et al. (1997) suggested a lifetime optimization methodology for planning the inspection and repair of structures that deteriorate over time and illustrated it through numerical examples. The optimization is based on minimizing the expected total life cycle cost while maintaining allowable lifetime reliability for the structure.

2.3.11 Risk-Informed Decision Making for Resilience Engineering

Risk management is typically based on the associated economics of design or improvement options identified for a particular decision situation based on the cost-effectiveness of an option for reducing the risk associated with the situation. As an example, in the context of extreme storms, options may include countermeasures aimed at reducing vulnerabilities of coastal lines, property, and asset exposure, impact on resources and populations, and land-use changes. Options may also include consequence mitigation strategies aimed at reducing the potential consequences, given the occurrence of a failure scenario (Ayyub 2014a). The probability of realizing a favorable benefit-to-cost ratio can be represented as follows:

$$P\left(\frac{\text{Benefit}}{\text{Cost}} \geq 1\right) = P(\text{Cost} - \text{Benefit} \leq 0) \tag{2-2}$$

where both benefit and cost in Equation (2-2) are random variables. With the knowledge of their underlying distributions, the probability of realizing a favorable benefit-to-cost ratio can be computed using reliability assessment techniques including Monte Carlo simulation. It should be noted that the uncertainty associated with the benefit is typically greater than the uncertainty associated with the cost of the strategy.

2.3.12 Community Socioeconomics

Examining community resilience, well-being, livability, and equity requires focusing on understanding the contribution of the economic, social, cultural, and political components of a community in maintaining itself and fulfilling the various needs of local residents (Kusel and Fortmann 1991). This examination builds on the very definitions of infrastructure and resilience. Losses are generalized in the framework as casualties, economic, social, and other things which sometimes are difficult to quantify, such as lower standard of living. This understanding is vital to knowing how to establish community performance targets. Although performance objectives related to the engineering are well established, linking them to the development of social objectives is challenging. The treatment of socioeconomic impacts is covered in Chapter 7, where the focus is strictly on resilience engineering.

Identifying target community performance objectives is essential to determining target infrastructure performance objectives. The performance objectives define measurable social and economic indicators of community resilience in terms of acceptable losses, including casualties, and time to return to social and economic function. The targets are developed relative to the hazard magnitudes and potential return periods the communities are exposed to. This establishes targets for infrastructure systems to meet in support of community resilience. This identification provides a link to supporting policy for improving community resilience through infrastructure systems.

Defining infrastructure system performance targets should be consistent with what is needed to meet the community performance objectives. Identifying target performance objectives in terms of lost performance, lost services, and time to rapidly restore them is essential to creating resilient infrastructure systems. Without target objectives, it is difficult to design resilience into the systems and impossible to predict whether the systems can provide the anticipated support for community resilience relative to the hazard exposure. System performance targets constitute the link to developing outcomes for the economics of resilience and regional social and economic losses. This includes establishing reasonable and affordable performance objectives.

The results of system performance are compared with target infrastructure system performance objectives. If the performance is better than the target performance objectives, then the system is considered to meet the expected resilience. If the performance is less than the target performance objectives, then modifications may be needed, possibly using advanced technologies or by resilience engineering, and the system reassessed. The social and economic and community performances with the community performance objectives may also need to be reevaluated. This comparison should be undertaken by engineers along with other sociology, economy, regional planning, and so on experts and is essential for ensuring that infrastructure systems perform resiliently in support of the community needs.

Resilient communities have the ability to appropriately manage without normal infrastructure services during the time they are lost, but resilience can be enhanced if there is some level of continuity of inoperable services, that is, the services that are temporarily lost. This requires the social units within communities to understand and integrate the technical and organizational resilience of infrastructure systems into their planning. Resilient infrastructure systems have the ability to accommodate service losses and restore them to the community when they are needed. This requires the infrastructure systems and the organizations operating them to understand and integrate the social and economic resilience of communities into their planning. Continuity of lost services can be achieved by using alternatives, substitution, maintaining stockpiles, going without, and other means (Rose 2016). These may be achieved by the community independent of the infrastructure systems, but to ensure resilience, planning for continuity of lost services should be accomplished collaboratively between the community and infrastructure system operators in coordination with government and nongovernment organizations that can provide aid until all services are returned. For example, when a water system loses services, the water system along with emergency managers and nongovernment responders can provide emergency access to bottled water. Loss of electric power may be supplemented with emergency generators. Loss of sewer system function may require access to portable toilets and showers. Rose (2016) described strategies useful for businesses.

2.4 PERFORMANCE TARGETS OF INFRASTRUCTURE SYSTEMS

Code-based performance standards are available, but they are not designed in coordination with the resilience objectives of a broader community. The standards focus on components or a single system but not a system of systems. The emphasis is on life safety with explicit consideration of potential losses of functionality in cases of extensive socioeconomic disruptions and slow recovery after a major hazard event. ASCE (2019) provides strategies and approaches to set targets for the performance of infrastructure systems necessary in the design and construction of buildings and lifeline systems to support a community's social stability, economic vitality, and environmental sustainability. It covers foundational topics including (1) characterization of communities and their supporting infrastructure, (2) hazard assessment, (3) metrics and assessment methods for community resilience, (4) metrics and assessment methods for building and lifeline performance, and (5) a conceptual framework for developing resilience-based performance standards.

2.5 INFORMATION AND DATA SOURCES

Information and data collection in support of all resilience elements are necessary. Some of the steps of the methodology might require information that is unavailable. In such cases, expert opinion elicitation and Bayesian methods could be employed (Ayyub 2001, 2002, 2014a). Information and data collection and sharing improve adaptive capacity, decision making, and situational awareness. Data sources are discussed in Chapter 1.

Assessing and managing infrastructure resilience can be based on Figure 2-1, requiring additional information on data on resilience governance, resilience-based performance standards and codes, loss estimation (Gilbert and Ayyub 2016, Ayyub et al. 2016), business continuity, emergency management, asset management, risk management (Ayyub 2014a), hazard scenarios, life-cycle cost assessment, decision support systems, and land use planning, to name a few.

2.6 EXAMPLES AND APPLICATIONS: TRANSPORTATION INFRASTRUCTURE

2.6.1 Introduction

This section presents an application of the methodology presented in Chapter 2 as a case study based on the response to the flooding and damage to Interstate-95 (I-95) in North Carolina following Hurricane Matthew in October 2016. This postevent case study highlights each step of the methodology, focusing on transportation infrastructure, more specifically, roads. The objective was to illustrate the application of each step of the methodology and to explore any insights provided by the application in terms of missing data and tools.

This section presents background on the case study and describes how the steps of system assessment and then governance and management apply to the case study.

2.6.2 Background and Methodology

I-95 runs from Maine to Florida along the eastern seaboard of the United States. This significant route is heavily traveled, carrying heavy volumes of passengers and truck traffic. The 292 km long (181 mi), four-lane wide section of I-95 in North Carolina diagonally crosses the eastern part of the state from the northern border with Virginia to the southern border with South Carolina. In 2015, average annual daily traffic (AADT) values vary along I-95 in North Carolina and are between 32,000 and 40,000 vehicles per day. The route is owned and maintained by North Carolina Department of Transportation (NCDOT). In early October 2016, rainfall from Hurricane Matthew caused the closure of I-95 in North Carolina. In addition to flooding, a major washout and bridge damage disrupted travel, requiring significant detours. Flooding and damage also closed many alternate routes. In this case study, we focus on I-95 and the alternate routes recommended by NCDOT. The data and models were originally assembled to explore measures of resilience (Liu et al. 2017). The models were then refined to provide a more rigorous assessment (Ren 2018).

The chronology of events shown in Table 2-1 was assembled from news reports and NCDOT news releases. The data are not intended to be

Table 2-1. Chronology of Events.

Date	Event	Assumptions Made for Analysis
October 7, 2016	Normal operations	Pre-event conditions
October 8, 2016	I-95 impassible at MM 44, MM 116, MM 119, and between Exit 25 and Exit 33	Trips are reduced by 10% and 20%; SB: All through traffic follows NCDOT suggested detours; some local traffic follows detours; most local traffic uses alternative routes; NB: Most traffic uses NCDOT suggested detours
October 10, 2016	I-95 impassible between Exit 13 and Exit 31	
October 12, 2016	I-95 SB closed between Exit 13 and Exit 56	
October 12, 2016	I-95 NB closed between Exit 13 and Exit 22	
October 15, 2016	I-95 open except for restrictions SB at MM 78	Users use one-lane section rather than alternate routes
October 18, 2016	All lanes open	Same as pre-event conditions

Note: MM = mile marker; NB = northbound; SB = southbound; and NCDOT = North Carolina Department of Transportation.

comprehensive or complete but to provide a realistic scenario. Detours suggested by NCDOT (2016) were used to develop a representative network, as shown in Figure 2-6. These detours and study areas are shown on the map in Figure 2-7. To understand the impact of I-95 closures, we model the diversion from I-95 to the routes shown in Figure 2-6 and represented in Figure 2-7. In Figure 2-7, Node 1 represents the junction of I-95 and US-64 (Exit 138), and Node 4 represents the junction of I-95 and I-74 (Exit 13). For southbound traffic, we use two intermediate nodes: Node 2 (Exit 119) and Node 3 (Exit 81). For northbound traffic, we only model the network between Node 4 (Exit 13) and Node 3 (Exit 81).

Each link is assigned attributes based on publicly available data from NCDOT, including AADT (from 2016), capacity, and free-flow travel time that are used to construct a link performance function based on the Bureau of Public Roads equation (Cambridge Systematics 2013a). An origin–destination matrix was inferred from link counts using an iterative process and verified using network equilibrium assignment (Ren 2018) using open-source software seSUE (Ahipaşaoğlu et al. 2019). The assumptions are documented in Liu et al. (2017).

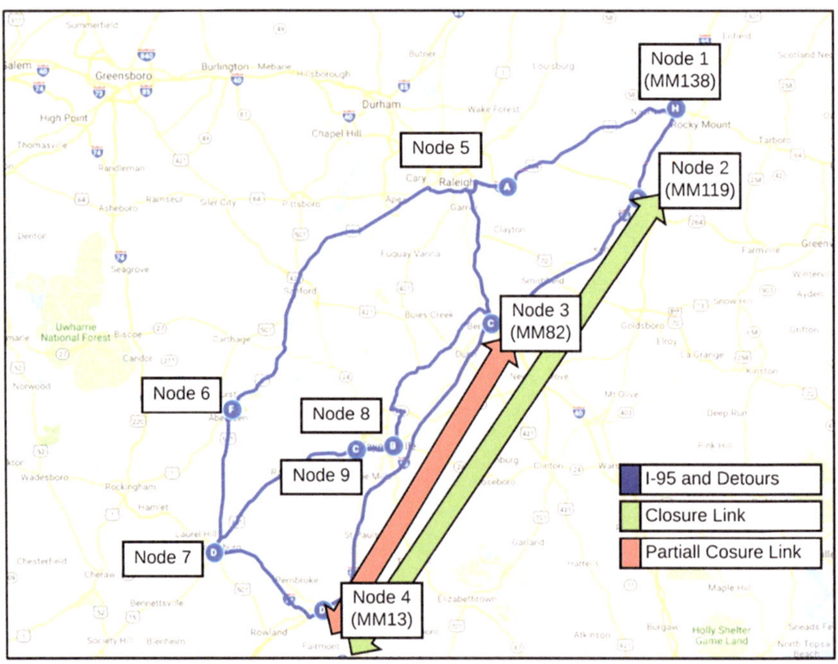

Figure 2-6. I-95 in North Carolina and Detours of Hurricane Matthew.
Source: Google Maps accessed October 12, 2019, and Ren (2018).

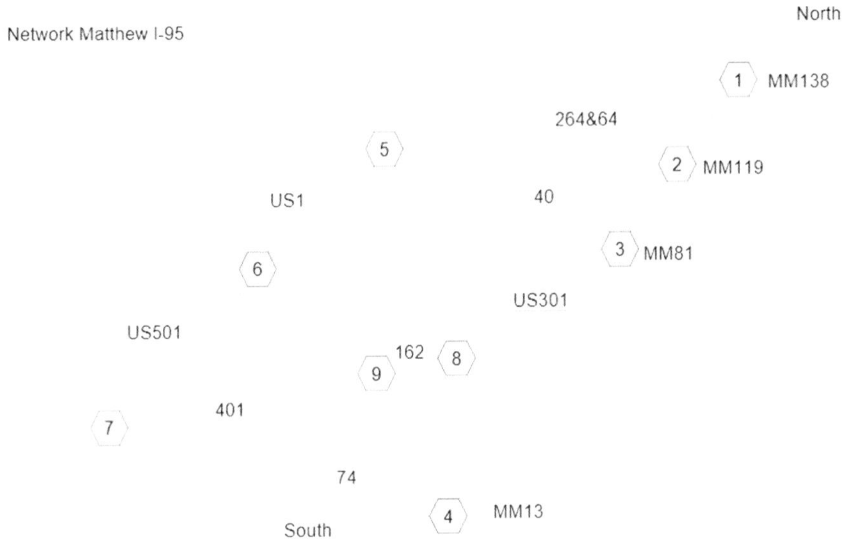

Figure 2-7. Simplified network of Hurricane Matthew.
Source: Ren 2018.

Several scenarios were developed to explore the sensitivity to various modeling assumptions, specifically

- 10% reduction in trips on I-95.
- 20% reduction in trips on I-95.
- 50% of traffic on I-95 uses the detour, and 50% of the traffic uses local routes, which are not modeled. Local routes are assumed to have a similar travel time to the direct route.

The models provided estimates of additional travel time and vehicle kilometers of travel.

2.6.3 System Assessment

2.6.3.1 Infrastructure Resilience Dimensions. The context, the first step of the methodology provided in Figure 2-1, is a multidimensional description of the system being considered. Each dimension and the appropriate description for this case study are presented in Table 2-2. This table then clearly specifies the system being discussed.

2.6.3.2 Transportation System Functionality. The transportation system functionality, as the next step of the methodology, specifies the performance measures elected to represent system functionality and how these measures vary over time. In this case, three measures representing functionality are

Table 2-2. Dimensions of the I-95 North Carolina Case Study.

Dimension		Case Study
Primary	Civil infrastructure or lifeline system	Transportation
	Hazard	Flooding
	Adverse event cycle	Recovery
Secondary	Resilience properties	Redundancy, rapidity, and resourcefulness
	Domain	Technical, social, and economic
	Geopolitical unit	State and regional

explored: availability measured as the ratio of capacity to the pre-event capacity; loss of mobility measured as the increase in vehicle kilometers traveled on the network; and accessibility measured as the additional travel time on the network.

The availability is shown in Figure 2-8 for the segments between Nodes 2 and 3 and Nodes 3 and 4 of the network. This measure applies only to the links damaged and does not capture the regional and statewide impacts that are recognized as the geopolitical unit as part of the domain identified in the first step of the methodology. Therefore, this is not an appropriate measure.

The measurement of either mobility or accessibility requires a network model. Initially, a simple network model was used (Liu et al. 2017). These models were subsequently refined, and the results for the additional travel time are shown in Figure 2-9 (Ren 2018). While much of the literature talks about the resilience triangle (Bruneau et al. 2003, see also Figure 2-3 with the *triangle* in this case is the performance drop represented by the red lines and recovery represented by the blue lines), the functionality of the transportation system is often a step function with abrupt changes when links are closed, partially open, or the service is restored. The greatest loss of functionality in the networked model in this case occurs over the time that the road was flooded and during the repair. If the entire network were modeled, the steps would be less significant and some functionality would be restored as flooded roads that are not damaged are reopened.

2.6.3.3 System Service Provision and Operability. The next (third) step of the methodology represents service provision and operability. The

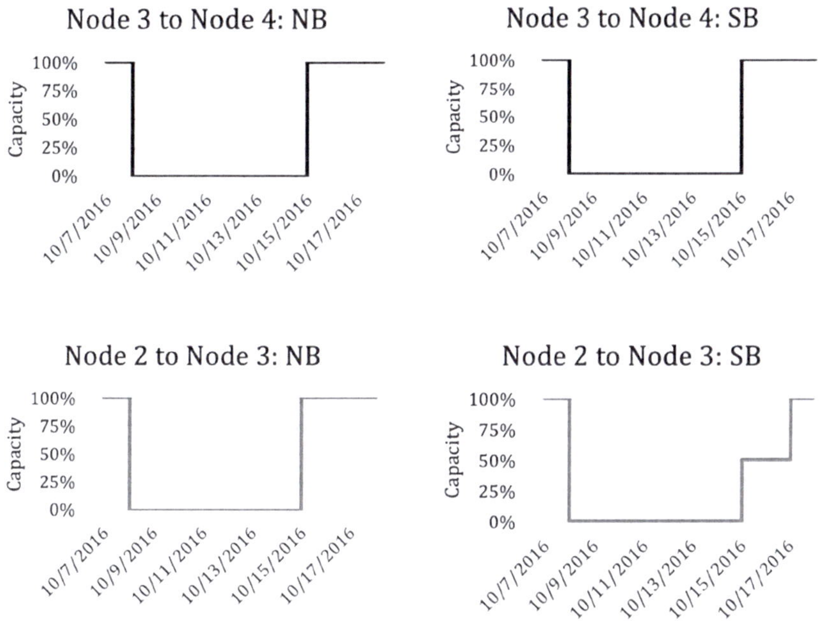

Figure 2-8. Functionality as measured by availability or capacity: (a) Node 3 to Node 4 NB, (b) Node 3 to Node 4 SB, (c) Node 2 to Node 3 NB, and (d) Node 2 to Node 3 SB.

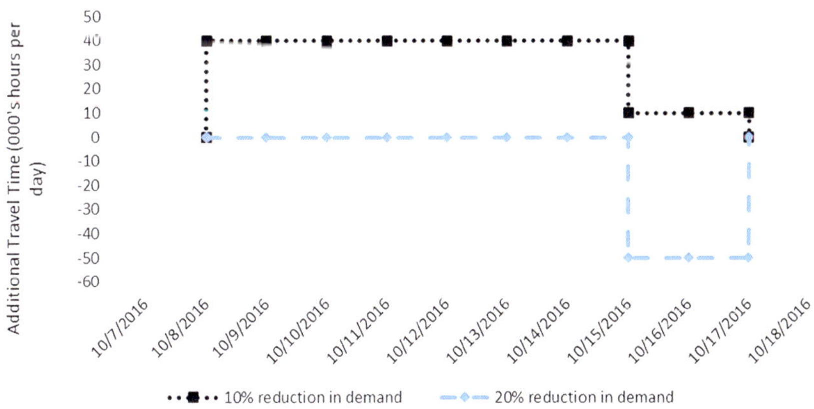

Figure 2-9. Accessibility over time.
Source: Ren (2018).

transportation system provides transportation services to the community. These services are often characterized as mobility, such as the movement of goods and people, and accessibility, such as the ability to reach goods, services, and destinations (Litman 2003). In addition to the measures of vehicle kilometers of travel and travel time presented in the previous subsection, mobility and accessibility can be measured using travel speeds, cost, or discomfort (Litman 2003). Measuring postevent performance or functionality as a percentage of pre-event performance, our models estimated the functionality over time, and made assumptions about traveler behavior. For example, Figure 2-9 shows the additional travel under two scenarios. The first assumes that the demand is reduced by 10% and the second reduced by 20%. These lost or incomplete trips represent a reduction in service provision that has to be captured. Hackl et al. (2018) assigned a value to each trip representing the cost of unsatisfied demand.

Figure 2-10 shows possible measures of service provision and operability, representing the percentage of transportation service delivered. Note that under partial closure and using the 10% reduction in trips on I-95, the travel time on the network is actually less than the pre-event conditions and more than 100% of transportation services (measures by excess travel time avoided) are delivered as there are travel time savings.

In the case of the I-95 closures in North Carolina, the network redundancy meant that alternative routes were available. While these routes were not necessarily convenient, they help to ensure the delivery of services.

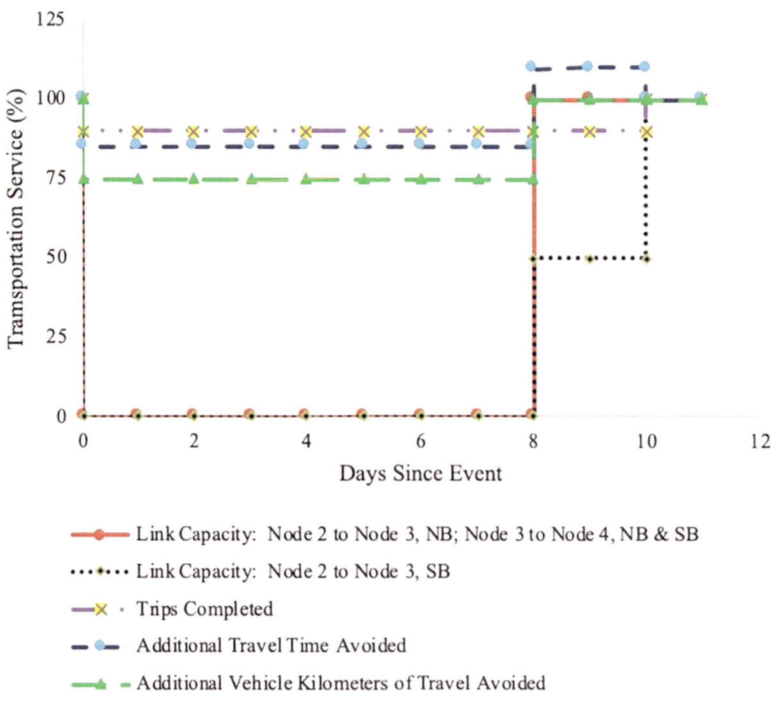

Figure 2-10. Percentage of transportation service delivered.

2.6.3.4 Continuity of Service Temporarily Lost. The fourth step of the methodology focuses on the resources available when some infrastructure components are inoperable. In this case, the transportation services provided by the inoperable links of I-95 in North Carolina were replaced by alternative routes including the detours specific in this analysis, local routes, and simply not making the trips. The relatively high level of redundancy in the transportation system supports the continuity of service.

2.6.3.5 Social and Economic Activity. This step of the methodology describes the social and economic activities of a community directly supported by the infrastructure system. The economic impacts can be monetized in terms of both direct and indirect costs [a distinction made by Hackl et al. (2018)], which are as follows:

- Direct costs: Costs to repair I-95 in Robeson County were approximately $800,000.
- Indirect costs
  - Additional travel time costs: Assuming $30/h for the value of travel time (Cambridge Systematics 2013a), the additional travel time is valued at $4.5 m.
  - Additional vehicle operating costs: Assuming $0.3125/km of travel, the additional cost of travel is $3.6 m.
  - Cost of lost trips: Using the Hackl value of a lost trip and the estimated 17,000 trips not made, over 10 days, the value of lost trips is $108/h; and each trip resulting in lost productivity of 8 h results in a cost of $18.4 m.

The estimates indicate that the direct costs ($800,000) are a small fraction of these indirect costs totaling $24.5 m for the 10-day closure. Other less tangible and measurable indirect costs include the following:

- Business disruption because of lack of accessibility (either directly because of damage or indirectly because of added congestion);
- Cost of disrupted supply chains;
- Deprivation costs owing to food, goods, and services being unavailable;
- Noneconomic loss such as casualties because of the inability to get to a hospital in time;
- Gains in the form of the increased remaining life of repaired roads, as shown in Figure 2-11; and
- Additional maintenance costs because of additional traffic on local roads and accelerated degradation owing to flooding, as illustrated in Figure 2-12.

The impacts on social activities include limited access to health care and other services, social and family interaction disruption, and equity issues.

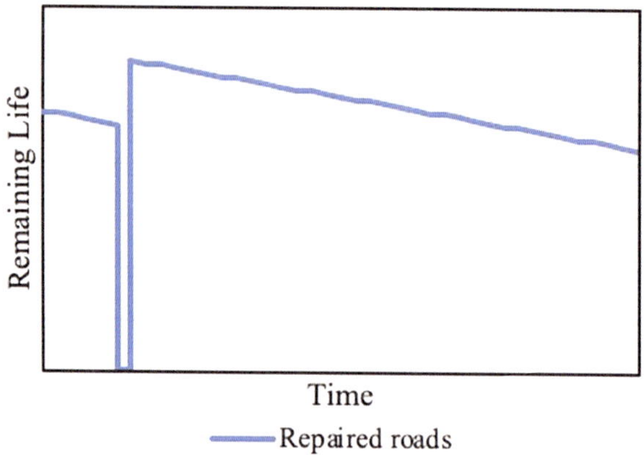

Figure 2-11. Remaining life gained by repaired roads.

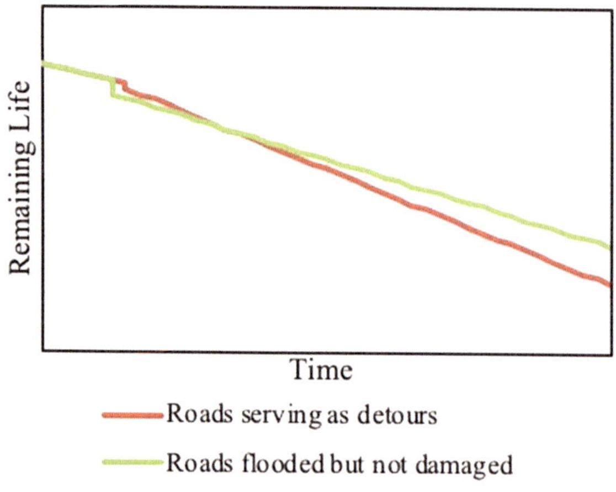

Figure 2-12. Changes in remaining life for other roads.

2.6.3.6 Community. This step of the methodology describes community resilience. Given the context (first step), it is difficult to distinguish the community resilience in the context of the I-95 disruptions from the local disruptions to the transportation system because of widespread flooding and the disruptions to other services such as education, health care, and government functions. Esnard et al. (2018) and van de Lindt et al. (2018) document the impacts of Hurricane Matthew on the community of Lumberton in Robeson, North Carolina, first with a focus on education, and second from the perspective of resilience. Specific community issues include stress and well-being, food and fuel shortages, the impacts of congestions and emissions on quality of life, and the disruption to other services in the right of way.

2.6.4 Governance and Management

The governance and management steps in the methodology are less connected to the context of this specific case study as this case study focuses on recovery, and such considerations are for a longer term. However, governance and management prior to Hurricane Matthew influenced the recovery and then the event itself; flooding because of Hurricane Matthew then influenced the governance and recovery.

Since Hurricane Matthew, North Carolina has had the unfortunate opportunity to test its resilience with respect to flooding twice more. In September 2018, flooding because of Hurricane Florence caused the closure of I-95. The chronology of events is summarized in Table 2-3. Given the extent of flooding, the detours were even longer and the disruption more extensive than those because of Hurricane Matthew flooding (Ren 2018).

In October 2018, Hurricane Michael again brought heavy rain and also the threat of tornados. Although local roads were closed, I-95 was not closed or damaged.

Table 2-3. Chronology of Events during Florence.

Start date	Event	Direction
September 14, 2018	Close I-95 at Exit 22	Heading north, left lane
	Close I-95 at Exit 72	Heading north, ramp close
September 15, 2018	Close I-95 Exit 19–22 and Exit 65–81	Both
	Close I-95 Exit 65–73	Heading north
	Close I-95 Exit 65–81	Heading south
	Close I-95 Exit 105	Heading south, right lane
September 16, 2018	Close I-95 Exit 13–22	Heading north
September 17, 2018	Close I-95 Exit 13–56	Both
September 18, 2018	Close I-95 Exit 18–20	Both
	Close I-95 at Exit 138	Heading south
September 24, 2018	All open	

Source: Ren (2018).

Two documents guide this discussion of governance and management as follows: North Carolina Department of Transportation (Transportation Asset Management Plan) (NCDOT 2018) and Hurricane Matthew Resilient Redevelopment Plan (Robeson County 2017).

2.6.4.1 Community Performance Targets. This seventh step is aimed at identifying community performance targets. In the case of roadways, this might be the number of days that a segment might be disrupted, given the functional classification. The North Carolina Resilient Redevelopment Planning (NCRPD) program is funded by the state legislature "to provide a roadmap for communities in eastern North Carolina to rebuild and revitalize after being damaged by Hurricane Matthew" (Robeson County 2017). However, no specific objectives or targets are specified.

The North Carolina Asset Management Plan (NCDOT 2018) explicitly identifies goals that include "Improve the reliability and connectivity of the transportation system," but the specific performance targets are focused on pavement and bridge conditions consistent with the requirements for such plans. The plan, as required by legislation (Cambridge Systematics 2013b), also includes a risk management analysis but that lacks specificity.

An important input to and outcome from this step is policy. The Asset Management Plan is the result of federal policy (in place pre-event), and the NCRPD program is the result of state policy postevent.

2.6.4.2 Infrastructure System Performance Targets. The eighth step is intended to provide the transportation system performance targets that support the community performance targets in Step 7. The postevent analysis of the closure of I-95 indicates that NCDOT's performance target was to restore the link as fast as possible.

2.6.4.3 Feedback. Comparing the infrastructure system performance targets (Step 8) to the performance over time (Step 2), the system service provision and operability (Step 3) and the continuity of service (Step 4) constitute an important feedback loop aimed at enhancing infrastructure resilience. In this postevent look at infrastructure, recovery provides an opportunity to determine what worked and what can be done better. Strategies such as regular communication to the public regarding closures, detour routes and openings, and partial opening of a segment aimed at enhancing resilience. To meet these targets, DOTs have mutual aid agreements, prepositioned equipment and materials, and capabilities for emergency contracting.

2.6.4.4 Economics and Resilience. Postevent planning (Robeson County 2017) is aimed at improving resilience. The NCRPD program focuses on not only just rebuilding but also revitalization. The Robeson County report lists "Upgrade Vulnerable Roads and Bridges" as a high priority and ranked this infrastructure pillar fourth. Unfortunately, the long lead time and high cost meant that these projects were not even in the planning stage when Hurricane Florence affected the region. The report also includes levee enhancements and enhanced flood warning systems that are likely to enhance the resilience of the communities as well as the transportation system.

2.6.4.5 Regional, Social, and Economic Losses. The regional, social, and economic losses include both direct and indirect gains and losses. While the importance of regional, social, and economic loss is acknowledged, the proposed lifecycle cost analysis in the Asset Management Plan (NCDOT 2018) focuses on direct costs. As a roadmap, the Robeson County (2017) plan provides a qualitative assessment. Therefore, there is much to be done in this area.

2.6.5 Observations and Conclusion

The methodology acknowledges the importance of the different perspectives and dimensions, the interdependencies among the different systems as they related to community resilience, and the underlying data and tools. Assessment of the resilience of the regional road network during disruptions from flooding and damage to I-95 required data on the event, alternative routes, network connectivity, link attributes, and travel demand. The analysis also relied on tools for network modeling and trip assignment.

These examples are intended to connect the steps of the resilience methodology using an actual event and using concepts from asset management in parallel with risk assessment and lifecycle cost analysis tools. This work underscored the importance of capturing all aspects of resilience and the need to enhance the tools supporting this kind of analysis. The examples underscore the importance of resilience. The direct costs are large and the indirect costs even greater.

REFERENCES

Ahipaşaoğlu, S. D., U. Arıkan, and K. Natarajan. 2019. "seSue." Accessed May 5, 2019. http://people.sutd.edu.sg/~ugur_arikan/seSue/.

ASCE. 2013.*Unified definitions for critical infrastructure resilience. Policy Statement 518.* Reston, VA: ASCE. Accessed October 20, 2018. http://www.asce.org/issues-and-advocacy/public-policy/policy-statement-518—unified-definitions-for-critical-infrastructure-resilience/.

ASCE. 2019. *Resilience-based performance: Next generation guidelines for buildings and lifeline standards.* Reston, VA: ASCE. Accessed January 8, 2020. https://sp360.asce.org/PersonifyEbusiness/Merchandise/Product-Details/productId/257999995.

Attoh-Okine, N. O., A. T. Cooper, and S. A. Mensah. 2009. "Formulation of resilience index of urban infrastructure using belief functions." *IEEE Syst. J.* 3 (2): 147–153.

Ayyub, B. M. 2001. "Uncertainties in expert-opinion elicitation for risk studies." In *Risk-based decision making in water resources IX*, edited by Y. Y. Haimes, D. A. Moser, E. Z. Stakhiv, G. Zisk, and B. Zisk, 103–121. Reston, VA: ASCE.

Ayyub, B. M. 2002. *Elicitation of expert opinions for uncertainty and risks.* Boca Raton, FL: CRC Press.

Ayyub, B. M. 2014a. *Risk analysis in engineering and economics.* 2nd ed. Boca Raton, FL: Chapman & Hall/CRC Press.

Ayyub, B. M. 2014b. "Systems resilience for multihazard environments: Definition, metrics, and valuation for decision making." *Risk Anal.* 34 (2): 340–355.

Ayyub, B. M. 2015. "Practical resilience metrics for planning, design, and decision making." *ASCE-ASME J. Risk Uncertainty Eng. Syst. Part A: Civ. Eng.* 1 (3): 04015008.

Ayyub, B. M., R. E. Chapman, G. E. Galloway, and R. N. Wright, eds. 2016. *Economics of community disaster resilience, workshop proceedings.* NIST Special Publication 1600. Gaithersburg, MD: National Institute of Standards and Technology, Office of Applied Economics. Accessed October 20, 2018. https://nvlpubs.nist.gov/nistpubs/SpecialPublications/NIST.SP.1600.pdf.

Ayyub, B. M., and G. J. Klir. 2006. *Uncertainty modeling and analysis in engineering and the sciences.* Boca Raton, FL: Chapman & Hall/CRC.

Ayyub, B. M., M. Medina, T. Vinson, D. Walker, R. N. Wright, A. Aghakouchak, et al. 2018. "Climate-resilient infrastructure: A manual of practice on adaptive design and risk management." In *Committee on adaptation to a changing climate*, edited by B. M. Ayyub. ASCE Manual of Practice Number 140. Reston, VA: ASCE. https://ascelibrary.org/doi/book/10.1061/9780784415191.

Biringer, B. E., E. D. Vugrin, and D. E. Warren. 2013. *Critical infrastructure system security and resiliency.* Boca Raton, FL: CRC Press.

Brown, N. 2018. "NYC steam system: Why we have it and how Con Ed maintains it." Accessed March 8, 2021. https://www.amny.com/news/nyc-steam-system-1-20146953/

Bruneau, M., S. E. Chang, R. T. Eguchi, G. C. Lee, T. D. O'Rourke, A. M. Reinhorn, et al. 2003. "A framework to quantitatively assess and enhance the seismic resilience of communities." *Earthquake Spectra* 19 (4): 733–752. https://doi.org/10.1193/1.1623497.

Cambridge Systematics. 2013a. "North Carolina I-95 economic assessment study, task 4: Travel demand modeling." Accessed January 8, 2020. https://connect.ncdot.gov/projects/Driving95/Travel%20Demand%20Modeling.pdf.

Cambridge Systematics. 2013b. *Generic work plan for developing a TAMP*. Washington, DC: US Dept. of Transportation, Federal Highway Administration. Accessed January 8, 2020. https://www.fhwa.dot.gov/asset/tamp/workplan.pdf.

Cimellaro, G. P., A. M. Reinhorn, and M. Bruneau. 2010. "Seismic resilience of a hospital system." *Struct. Infrastruct. Eng.* 6 (1–2): 127–144.

conEdison. 2018. "Steam service." Accessed September 4, 2018. https://www.coned.com/en/commercial-industrial/steam.

Davis, C. A. 2013. "Quantifying post-earthquake potable water system services." In *6th China–Japan–US Trilateral Symposium Lifeline Earthquake Engineering*, 19–26. TCLEE Monograph No. 38. Reston, VA: ASCE.

Davis, C. A. 2014. "Water system service categories, post-earthquake interaction, and restoration strategies." *Earthquake Spectra* 30 (4): 1487–1509.

Davis, C. A. 2019. "Infrastructure system resilience: Functionality and operability." In *Proc., 2nd Int. Conf. on Natural Hazards and Infrastructure*. June 23 – 26, 2019, Chania, Greece Paper No. 200, Accessed March 8, 2021 https://iconhic.com/2019/

Davis, C. A., A. Mostafavi, and H. Wang. 2018. "Establishing characteristics to operationalize resilience for lifeline systems." *Nat. Hazard. Rev.* 19 (4): 04018014.

Der Kiureghiana, A., and O. Ditlevsen. 2009. "Aleatory or epistemic? Does it matter?" *Struct. Saf.* 31 (2): 105–112.

Ellingwood, B. R., and J. Y. Lee. 2016. "Life cycle performance goals for civil infrastructure: Intergenerational risk-informed decisions." *Struct. Infrastruct. Eng.* 12 (7): 822–829.

Esnard, A.-M., B. S. Lai, C. Wyczalkowski, N. Malmin, and H. J. Shah. 2018. "School vulnerability to disaster: Examination of school closure, demographic, and exposure factors in Hurricane Ike's wind swath." *Nat. Hazard.* 90 (2): 513–535.

Frangopol, D. M., K.-Y. Lin, and A. C. Estes. 1997. "Life-cycle cost design of deteriorating structures." *J. Struct. Eng.* 123 (10): 1390–1401.

Gilbert, S., and B. M. Ayyub. 2016. "Models for the economics of resilience." *ASCE-ASME J. Risk Uncertainty Eng. Syst. Part A: Civ. Eng.* 2 (4). Accessed March 8, 2021. https://ascelibrary.org/doi/10.1061/AJRUA6.0000867

Hackl, J., J. C. Lam, M. Heitzler, B. T. Adey, and L. Hurni. 2018. "Estimating the risk related to networks: A methodology and an application on a road network." *J. Nat. Hazards Earth Syst. Sci.* 18 (9): 2273–2293.

Holling, C. S. 1973. "Resilience and stability of ecological systems." *Annu. Rev. Ecol. Syst.* 4 (1): 1–23.

Kaplan, S., and B. J. Garrick. 1981. "On the quantitative definition of risk." *Risk Anal.* 1 (1): 11–27.

Kusel, J., and L. P. Fortmann. 1991. "What is community well-being?" In Vol. 1 of *Well-being in forest-dependent communities*, edited by J. Kusel and L. Fortman, 1–45. Berkeley, CA: Forest and Rangeland Resources Assessment Program and California Dept. of Forestry and Fire Protection.

Litman, T. 2003. "Measuring transportation: Traffic, mobility and accessibility." *Inst. Transp. Eng. ITE J.* 73 (10): 28–32.

Liu, Y., S. McNeil, and R. Lee. 2017. "Operationalizing the concept of resilience: A case study of flooding in North Carolina." In *Proc., Maintenance and Rehabilitation of Constructed Infrastructure Facilities*. Organized by International Society for Maintenance and Rehabilitation of Transportation Infrastructures (iSMARTi), Seoul, Korea. Accessed March 8, 2021. https://www.concretepavements.org/event/2017-maireinfra-international-conference-on-maintenance-and-rehabilitation-of-constructed-infrastructure-facilities-in-seoul/

NCDOT (North Carolina Department of Transportation). 2016. "Sec. Tennyson visits areas flooded by Matthew." Press Release. Accessed January 8, 2020. https://www.ncdot.gov/news/press-releases/Pages/2016/Sec-Tennyson-Visits-Areas-Flooded-by-Mat.aspx.

NCDOT. 2018. "Transportation asset management plan." 2018 Interim Report. Accessed January 8, 2020. http://www.tamptemplate.org/wp-content/uploads/tamps/056_northcarolinadot.pdf.

NIST (National Institute of Standards and Technologies). 2015. *Community resilience planning guide for buildings and infrastructure systems*. NIST Special Publication 1190. Gaithersburg, MD: NIST.

NRC (National Research Council). 2012. *Disaster resilience: A national imperative*. Washington, DC: National Academies Press.

PPD (Presidential Policy Directive). 2013. *Critical infrastructure security and resilience*. PPD-21. Washington, DC: PPD. Accessed October 20, 2018. https://www.dhs.gov/sites/default/files/publications/ISC-PPD-21-Implementation-White-Paper-2015-508.pdf.

Ren, T. 2018. "Operationalizing the concept of resilience: Case study of flooding in North Carolina." *Senior thesis*, Dept. of Civil and Environmental Engineering, University of Delaware.

Robeson County. 2017. "Hurricane Matthew resilient redevelopment plan." Accessed January 8, 2020. https://files.nc.gov/rebuildnc/documents/matthew/rebuildnc_robeson_plan_combined.pdf.

Rose, A. 2016. *Benefit–cost analysis of economic resilience actions*. Oxford, MS: Oxford Research Encyclopedia of Natural Hazard Science, Oxford University Press.

Shinozuka, M., M. Q. Feng, J. Lee, and T. Naganuma. 2000. "Statistical analysis of fragility curves." *J. Eng. Mech.* 126 (12): 1224–1231.

Tierney, K., and M. Bruneau. 2007. "Conceptualized and measuring resilience." *TR News* 250: 14–17. Accessed April 8, 2015. http://onlinepubs.trb.org/onlinepubs/trnews/trnews250_p14-17.pdf.

UNISDR (United Nations Office for Disaster Risk Reduction). 2012. *Making cities resilient: My city is getting ready! A global snapshot of how local governments reduce disaster risk.* Geneva: UNISDR. Accessed October 20, 2018. www.unisdr.org/campaign.

van de Lindt, J. W., W. G. Peacock, J. Mitrani-Reiser, N. Rosenheim, D. Deniz, M. K. Dillard, et al. eds. 2018. *The Lumberton, North Carolina flood of 2016: A community resilience focused technical investigation.* Special Publication (NIST SP)-1230. Gaithersburg, MD: NIST. Accessed January 8, 2020. https://nvlpubs.nist.gov/nistpubs/SpecialPublications/NIST.SP.1230.pdf.

Van der Merwe, S. E., R. Biggs, and R. Preiser. 2018. "A framework for conceptualizing and assessing the resilience of essential services produced by socio-technical systems." *Ecol. Soc.* 23 (2): 12.

CHAPTER 3
RESILIENCE ASSESSMENT METHODS

3.1 BACKGROUND: UNCERTAINTY AND RISK

Uncertainties in hazards affecting infrastructure systems pose challenges to engineers and pragmatically require developing and using appropriate approaches. Chapter 2 provides some details on uncertainty types in terms of aleatory or epistemic types. In addition, uncertainties associated with policy and decision making are fundamentally attributable to the associated risks. Risk methods provide practical means for dealing and managing uncertainty (Ayyub 2014b). It is commonly measured in simple terms as the probability of the occurrence of an event or a scenario of events and the (unfavorable) outcomes or consequences associated with the occurrence of an event (stressor). Risk assessment is primarily concerned with answering three questions, as discussed in Chapter 3 of Kaplan and Garrick (1981): (1) What could happen, that is, what could go wrong? (2) How likely is it to happen? (3) If it does happen, what are the consequences? Risk assessment is a systematic process to identify the potential uncertain events including hazards to determine the consequences of event occurrences and to estimate its occurrence likelihood.

Risk reflects the probability of an unacceptable consequence because of the occurrence of natural or human-caused hazards. In general, neither the demand nor the available capacity is known with certainty, that is, they are not deterministic.

For a given system or component design, one can plot the available capacity against the probability of realizing it. In general, normal or lognormal probability distributions are suitable for the practical representation of the capacity. Time dependency can be introduced to represent aging and degradation. As one would expect, characterization of the system capacity involves a lot more uncertainty than that for characterization of the capacity of constituent components.

For hazards, a plot of the magnitude of each hazard type and the associated frequency of occurrence can be constructed (referred to as hazard curve); however, the hazard process is typically treated as stationary, although it can be nonstationary because of a changing climate as an example; that is, the hazard curve reflects projection based on currently available data. It is noted that the demand for a system or component (i.e., required capacity) because of a given hazard intensity level is also uncertain because of the many modeling and other uncertainties associated with demand analysis, and accordingly, the required capacity should be defined using appropriate statistical and analytical models.

A fragility curve can be used to represent the conditional probability of failure associated with the occurrence of a hazard of a particular intensity (i.e., the conditional probability that the associated required capacity exceeding the available capacity). Reliability is then obtained by convolving the hazard curve with the probability distribution for available capacity.

3.2 RESILIENCE ASSESSMENT AND QUANTIFICATION SCOPE: MODELS AND METHODS

In general, distinctions among resilience assessment methods are typically based on the following attributes:

- Quantitative versus qualitative assessment,
- Deterministic versus probabilistic methods,
- Components versus systems, and
- Networks versus systems of systems.

On a fundamental level, many resilience definitions include mechanisms by which the infrastructure responds to the changes, such as

- Ability to absorb or withstand the impact of the change;
- Ability to adapt in response to the change;
- Ability to recover and restore system functionality rapidly; and
- Efficiency or amount of resources required to successfully respond to a disruption, such as manpower, equipment, and other critical resources for response and recovery operations.

Resilience assessment methods can be conveniently grouped as follows: (1) attribute-based methods, and (2) performance-based methods. Attribute-based methods generally seek to answer the question "What makes my system more (or less) resilient?" Thus, these methods typically include system properties that are accepted as being beneficial to resilience. Examples of these categories might include robustness, resourcefulness, adaptivity, and recoverability. Application of these methods typically requires analysts to follow a process to review their system and determine the degree to which the properties are present within the system. The benefit of these approaches is that their applications tend to be less time- and resource-intensive and result in either qualitative or semiquantitative estimates of resilience. However, these approaches do not provide any estimation or confidence in how well the system will operate in the event of a disruption or the effectiveness of potential resilience enhancements and investments. The Supply Chain Resilience Assessment and Management tool (Petit et al. 2010) and Argonne National Laboratory's Resilience Index (Fisher and Norman 2010) are two examples of attribute-based methods. In general, performance-based methods are quantitative methods that try to answer the question "How resilient is my system?" These methods are used to interpret quantitative data that describe infrastructure outputs and to formulate metrics of infrastructure resilience. The required data can be gathered from historical events, subject–matter estimates, or computational infrastructure models. These methods tend to rely less on subjective or qualitative evaluations and, thus, facilitate comparative analyses. Because the metrics can be used to measure the potential benefits and costs associated with proposed resilience enhancements and investments, performance-based methods are often ideal for cost–benefit and planning analyses. Performance-based methods often use computational models to generate the necessary data, and these models may require significant time and resources to develop. Consequently, performance-based methods can be comparatively more complex. When deciding which methods to use, the analyst should determine their analysis objectives, evaluate their resources for performing the analysis, and assess their comfort with the varying levels of complexity. Assessment approaches were developed by the Multidisciplinary Center for Earthquake Engineering Research (Bruneau et al. 2003), and Rose (2007) and Alderson et al. (2015) developed examples of performance-based methods. Attribute- and performance-based approaches have their benefits and limitations, but when jointly considered, they have the potential to inform infrastructure stakeholders with a more complete understanding of infrastructure resilience. That is, they can describe not only "How resilient is my system?" but also "What can I do to make my system more resilient?"

In addition, resilience methods may include complex networks of systems, agent-based consideration to represent users, control and approximate reasoning including fuzzy logic. These methods might focus on a particular component of an adverse-event cycle starting from a hazard to recovery, as represented in Figure 2-2, such as resist, recover, stabilize, rebuild, and adapt. The type of disruptive event, such as natural or human caused, might necessitate the appropriate method to use. The type of infrastructure or system (such as electric power, natural gas, transportation, water, and so on) is a primary consideration in selecting methods. The scale (such as a single project or a system of projects, or local versus regional or statewide or national or global), and the relation to sustainability, among other considerations, might have a great influence on selecting methods. An engineer or analyst should make an informed decision regarding the level of complexity warranted depending on project needs in the light of these considerations.

Many resilience assessment models and methods are available in the literature. This chapter starts with three introducing fundamental resilience quantification models: (1) a resilience-triangle, (2) an availability-based resilience, and (3) a simplified model. These fundamental models provide a foundation for resilience quantification methods and are covered in Section 3.3.

With regard to scale, resilience assessment of a single facility or system is relatively manageable compared with that of complex interdependent systems or networks. Analyzing a facility, however, requires drawing boundaries around it and making appropriate assumptions regarding its interdependencies with its connected systems or networks. Resilience assessment of systems of systems or networks requires other system-centric concepts and approaches beyond those presented in Section 3.3. The following four holistic resilience assessment methods are discussed in this chapter:

1. Resilience assessment for a single facility (Section 3.4),
2. Resilience assessment of systems using a simplified method (Section 3.5),
3. Resilience assessment of systems of systems using a practical framework (Section 3.6), and
4. Resilience assessment of infrastructure networks (Section 3.7).

Section 3.4 focuses on a single system or facility, called entity, because of its relative simplicity. Resiliency assessment of such entities can be addressed using the fundamental concepts and models introduced in Section 3.3, along with additional concepts that can be specific to the type of system or facility and associated hazards.

Section 3.5 presents a method called infrastructure resilience analysis method (IRAM) for simplifying the analysis of systems. It is an analytical method for quantifying resilience based on two elements: systemic impact and total recovery efforts. It is a hybrid approach that combines performance-based metrics with attribute-focused analysis to address the questions of "How resilient is my system?" and "What can I do to make my system more resilient?" A resilience assessment example for a freight rail network subjected to a postulated flooding scenario is presented. This method is suitable for simplifying the treatment of systems.

Section 3.6 introduces system-of-systems (SoS) methods for resilience assessment of infrastructure systems. A method is suggested in terms of a framework for specifying, characterizing, and modeling different human and physical system components and their interactions at various analysis levels. Two case studies are provided. The first case study applies the SoS approach for assessing the resilience of a transportation network to the impacts of long-term sea-level rise in coastal areas. The second case study is related to evaluation of the effects of population fluctuation and funding gaps on the long-term resilience of a water distribution network.

Section 3.7 provides methods for assessing the topological vulnerability and resilience of infrastructure networks. A nonhazard-specific network vulnerability assessment example for a metro (i.e., subway) network is provided. In this section, another resiliency assessment method focusing on recovery optimization is presented with an example of an electric power infrastructure network subjected to a seismic.

3.3 FUNDAMENTAL MODELS FOR QUANTIFYING RESILIENCE

Several definitions of infrastructure resilience were discussed in Chapter 2. Among them is the definition by the Presidential Policy Directive (PPD)-8 (PPD 2011) as "the ability to adapt to changing conditions and withstand and rapidly recover from disruption due to emergencies." Measuring resilience is typically based on the performance of an infrastructure system after an external shock including the time it takes to return to initial level of performance. This section provides several fundamental methods for quantifying resilience.

3.3.1 Resilience-Triangle Model

The fundamental concept of resilience is illustrated in Figure 3-1 of performance as a function of time. The figure shows what is termed as the resilience triangle that can be used as a basis to compute a resilience index defined on a fundamental level as follows:

$$\text{Resilience} = \frac{\int_{t_0}^{t_1} Q(t)dt}{100(t_0 - t_1)} \tag{3-1}$$

where

Q = Infrastructure performance (or quality),

t_0 = Time of incident (or disturbance) occurrence, and

t_1 = Time to full recovery.

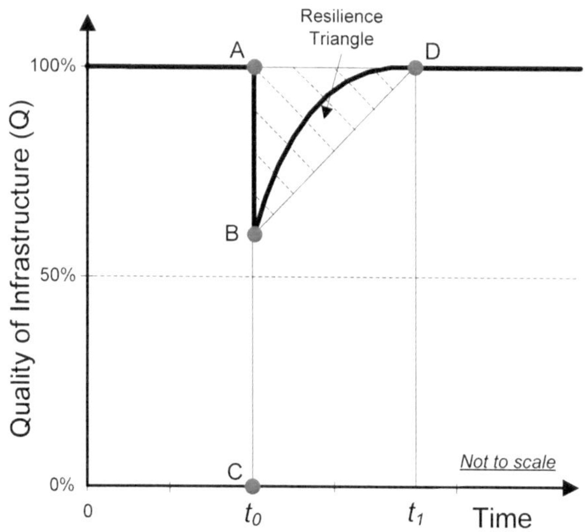

Figure 3-1. Resilience properties and triangle.

The performance of a system can be viewed as its functionality. The treatment of interdependencies across physical infrastructure systems requires examining the return of the interdependent systems to functionality, creating complexity addressed in later sections of this chapter, such as networks and system of systems necessary to support societal needs, as discussed in Chapter 6.

Resilience according to Equation (3-1) is dimensionless, where performance (Q) representing infrastructure quality according to Figure 3-1 can be measured in percentage or some native units if necessary. The earthquake community used Equation (3-1) with a suggested approach of resilience called the four Rs defined as follows (Bruneau et al. 2003):

- Robustness as the ability of the system and system elements to withstand external shocks without significant loss of performance;
- Redundancy as the ability of a system to function even after some of its components or subsystems have failed or more broadly the extent to which the system satisfies and sustains functional requirements after the failure of some of its components or subsystems;
- Resourcefulness as the ability to diagnose and prioritize problems and to initiate solutions by identifying and monitoring all resources, including economic, technical, and social information; and
- Rapidity as the ability to recover and contain losses and avoid future disruptions.

Table 3-1. Definition of Resilience Properties.

Property	Models (Points A, B, C, and D according to Figure 3-1)	Units	
Robustness	Robustness = B – C	Percentage	(3-2)
Redundancy	Not defined		
Resourcefulness	Not defined		
Rapidity	Rapidity $= \dfrac{A - B}{t_1 - t_0}$	Average recovery rate in percentage per time	(3-3)

These properties can be quantified as defined in Table 3-1 that makes reference to Figure 3-1.

3.3.2 Availability-Based Resilience Model

Another model provided by Ayyub (2014a) is illustrated in a schematic representation of a system's performance (Q) with aging effects and an incident occurrence with a rate (λ) according to a Poisson process in Figure 3-2. At time t_i, it might lead to degraded performance, called failure for convenience, of a duration ΔT_f. The failure event concludes at time t_f. The failure event is followed by a recovery event with a duration ΔT_r. The recovery event concludes at time t_r. The total disruption (D) has a duration of $\Delta T_d = \Delta T_f + \Delta T_r$. For illustration purposes the figure shows three failure events, brittle (f_1), ductile (f_2), and graceful (f_3); and six recovery events, expeditious recovery to better than new (r_1), expeditious recovery to as good as new (r_2), expeditious recovery to better than old (r_3), expeditious recovery to as good as old (r_4), recovery to as good as old (r_5), and recovery to worse than old (r_6). These events define various rates of change of performance of the system. The figure also shows the aging performance trajectory and the estimated trajectory after recovery. Resilience can be measured by focusing on the timescale in a manner that is consistent with the concept of availability used in reliability engineering as follows:

$$\text{Resilience}\,(R_e) = \frac{T_i + F\Delta T_f + R\Delta T_r}{T_i + \Delta T_f + \Delta T_r} \tag{3-2}$$

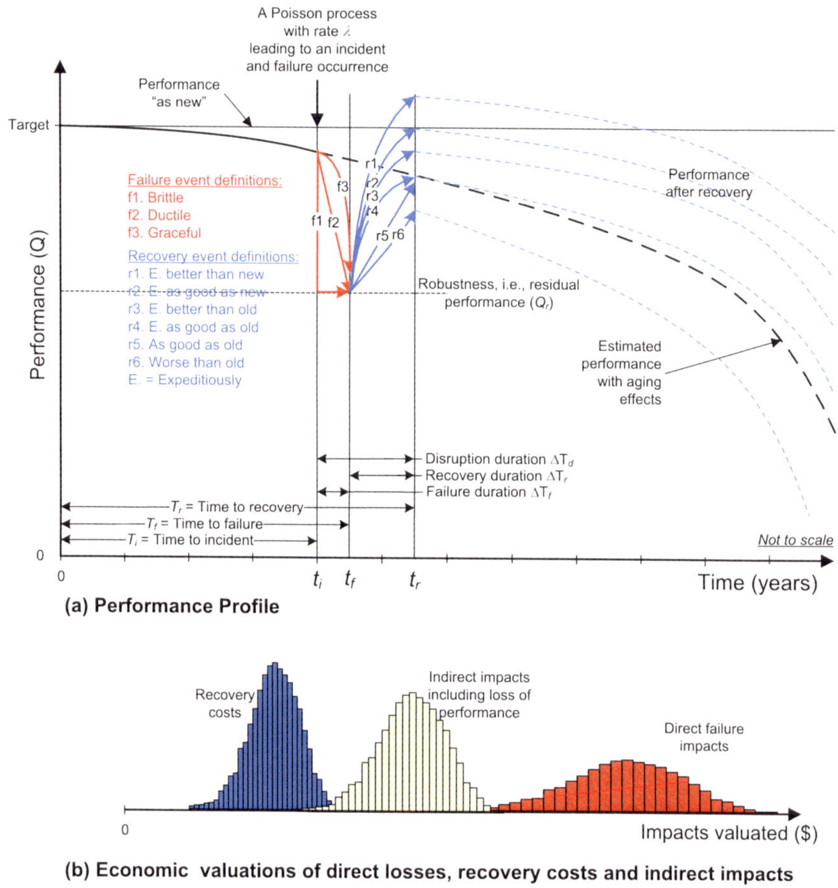

Figure 3-2. Definitions of resilience metrics: (a) performance profile, and (b) economic valuations of direct losses, recovery costs, and indirect impacts. Source: Ayyub (2014a).

where for any failure event represented by curve f as illustrated in Figure 3-2 by f_1, f_2, and f_3 red curves, and the corresponding *failure profile* (F) is measured as follows:

$$\text{Failure}(F) = \frac{\int_{t_i}^{t_f} f dt}{\int_{t_i}^{t_f} Q dt} \tag{3-3}$$

The integral in the numerator is of the failure curve f over time from t_i to t_f, whereas the integral in the denominator is of the varying performance curve Q over time from t_i to t_f. Similarly, for any recovery event represented by curve r as illustrated in Figure 3-2 by r_1 to r_6 blue curves, the corresponding *recovery profile* (R) is measured as follows:

$$\text{Recovery}(R) = \frac{\int_{t_f}^{t_r} r dt}{\int_{t_f}^{t_r} Q dt} \tag{3-4}$$

In this case, the integral in the numerator is of the recovery curve r over time from t_f to t_r, whereas the integral in the denominator is of the varying performance curve Q over time from t_f to t_r. The failure profile value (F) can be considered as a measure of robustness and redundancy and proposed to address the notion offered by Equation (3-3), whereas the recovery profile value (R) can be considered as a measure of resourcefulness and rapidity and proposed to address the notion offered by Equation (3-4). Equation (3-2) measures resiliency as a weighted availability by the corresponding failure and recovery profiles F and R, respectively.

Building on the work of Mori and Ellingwood (1993), the time to failure (T_f) can be characterized by its probability density function computed as follows for a particular component of a system:

$$-\frac{d}{dt}\int_{s=0}^{\infty} \exp\left[-\lambda t\left(1-\frac{1}{t}\int_{\tau=0}^{t} F_L(\alpha(\tau)s)d\tau\right)\right] f_{S_0}(s)ds \tag{3-5}$$

where Q is defined as the system's performance in terms of its strength (S) minus the corresponding load effect (L) in consistent units, that is, $Q = S - L$. Both L and S are treated as random variables, with F_L being the cumulative probability distribution function of L and f_S being probability density function of S. The aging effects are considered in this model by the term $\alpha(t)$ representing a degradation mechanism as a function of time t. It should be noted that the term $\alpha(t)$ can also represent improvement to the system. Equation (3-5) is based on a Poisson process with an incident occurrence, such as loading, rate of λ. The probability density function of T_f as shown in Equation (3-5) is the negative of the derivative of the reliability function. Times T_i, T_f, and T_r are random variables, as shown in Figure 3-2 and related to durations as follows:

$$\Delta T_f = T_f - T_i \tag{3-6}$$

$$\Delta T_r = T_r - T_f \tag{3-7}$$

The disruption duration is given by

$$\Delta T_d = \Delta T_f + \Delta T_r \tag{3-8}$$

Relating the time-varying reliability of components in accordance with Equation (3-5) to the reliability of a system requires using system reliability concepts as provided in textbooks such as by Ayyub (2014b).

Figure 3-2 also shows the associated costs including losses, recovery costs, and indirect costs that can be treated using methods described in Chapter 4. These losses and costs should be based on total economic valuations using anthropocentric considerations.

3.3.3 Simplified Resilience Model

A simplified model (Ayyub 2015) is available for a fundamental case of having a performance level that would be maintained and sustained over time, that is, no aging effects, with a brittle failure profile, that is, f_1, provided in Figure 3-2. Also, assume that the recovery is *as good as old* (i.e., r5 in Figure 3-2). In addition, the following assumptions are made: (1) a planning horizon (t), (2) a Poisson process of stressors with a rate (λ), (3) the planning horizon related to the stressor rate as $t = 1/\lambda$, (4) failure probability (p) because of a stressor, and (5) independent failures. It should be noted that the stressors have varied intensities and not all stressors fail the system and disrupt the system's performance. Fundamental cases are presented in this section: the case of linear recovery and the case of step recovery; however, other cases can be treated in a similar manner.

For the fundamental case of linear recovery, the resilience metric of Equation (3-2) for one failure-causing event is basically the ratio of two areas according to this figure, that is, the rectangular area tQ_{100}, divided by the tQ_{100} without the triangle representing the degraded performance of the system. The triangle has the sides of brittle failure and linear recovery. For a linear recovery path (r), it can be expressed as follows for one failure-inducing event (Ayyub 2015):

$$\text{Linear recovery:}\quad \text{Resilience per failure}(R_f) = 1 - \frac{(t_r - t_i)(Q_{100} - Q_r)}{2Q_{100}t} \tag{3-9}$$

For analytical and computational convenience, the concept of nonresilience, that is, resilience triangle, can be introduced and defined as follows:

$$\text{Linear recovery:}\quad \text{Nonresilience per failure}(\bar{R}_f) = \frac{(t_r - t_i)(Q_{100} - Q_r)}{2Q_{100}t} \tag{3-10}$$

The relationship between R_f and $\bar{R}_f$ is

$$\bar{R}_f = 1 - R_f \tag{3-11}$$

Equations (3-11) and (3-12) can be generalized to account for the potential of multiple occurrences of failure-inducing events x and their associated probabilities as follows:

$$\text{Resilience } (R_e) = 1 - \sum_{x=1}^{\infty}\left(\exp(-\lambda t)\frac{(\lambda t)^x}{x!}p^x\bar{R}_f^x\right) \tag{3-12}$$

This equation can be reduced to

$$\text{Resilience } (R_e) = 1 - \exp\left(-\lambda t\left(1 - p\bar{R}_f\right)\right) + \exp(-\lambda t) \tag{3-13}$$

This model offers the simplicity and practicality desired for systems with time-invariant performance and accounts for the following: rate of events, that is, the rate λ of a Poisson process representing stressor occurrences; conditional probability of failure (p) given or under the condition of the occurrence of a stressor based on the strength of the system; capacity of the system (Q_{100}); robustness of the system (Q_r); brittle failure and linear or step recovery to as-good-as-old profiles; nonresilience associated with the occurrence of a failure-inducing event; and planning horizon t. Ayyub (2015) provides the derivation of underlying equations for measuring resilience according to this simplified model.

3.4 RESILIENCE ASSESSMENT OF A SINGLE SYSTEM OR FACILITY

3.4.1 Selected Methods for a Single System or Facility

In general, a single system (or facility) is rarely self-contained and self-reliant, although this is theoretically achievable and may even be necessary for certain mission-critical applications. Accordingly, resilience assessment of a single system/facility requires making assumptions regarding the resilience aspects of its supporting and dependent systems or networks. In any case, this approach is often a relevant exercise for resilience assessment of complex networks (i.e., interdependent systems) because their resilience depends on accurate resilience analysis of the constituent systems or facilities. Therefore, it is important to have a good grasp of resilience assessment of the single system or facility case (i.e., analogous to how it is necessary to perform fragility analysis for all constituent structures, subsystems, and components to conduct accurate probabilistic risk analysis of a complex system). This aspect can also be appreciated from the North Carolina transportation infrastructure example covered in Section 2.6.

Most complex networks comprise many individual systems or facilities. For example, a health-care network consists of individual hospitals, outpatient centers, and numerous supporting facilities such as pathology and radiology labs. A transportation network consists of bridges, highways, local/regional roads, railroads, subway systems, individual train/subway stations, and so on. Similarly, a water supply network consists of lakes/reservoirs/rivers, water purification centers, water distribution pipelines, and so on. The constituent facilities/components can be fairly complex and possess their own resilience characteristics. Any such single system/facility could be the weakest link in terms of the overall resiliency of the associated network, and hence, it is, in general, important or desirable to perform accurate resilience analysis/assessment.

Depending on their relative contribution to the network's overall resilience, informed decisions can be made regarding the need for targeted upgrades of individual systems or facilities. On the contrary, depending on the safety or operability significance of the individual system/facility, it may be necessary to introduce some measures of self-reliance (i.e., network independence), whereby the pertinent dependency on the associated network(s) is reduced or eliminated. An example of this approach is the recent US-wide implementation of diverse and flexible coping strategies (collectively referred to as *FLEX* strategies) that ensure that nuclear power plants have access to extended back-up power and cooling capability to prevent the risk of Fukushima-like nuclear accidents (Powell et al. 2013). The need for introducing such self-reliance feature(s) obviously depends on the criticality of the individual system or facility, and its contribution to the vulnerability of an underlying network.

Models covered in Section 3.3 are generally adequate for performing resilience assessment of individual systems or facilities. However, depending on its role or function and the nature of the associated network, each type of individual system/facility requires consideration of relevant parameters (e.g., a medical facility). Sometimes a domain-specific or hazard-specific tool [e.g., the BuroHappold Resilience Framework as provided by Field et al. (2017) or a screening tool such as Resilience-based Earthquake Design Rating System according to Almufti and Willford (2014)] can be utilized to score an existing facility and to determine its upgrade needs. Such tools could also be used for the design of a new facility. Chapter 5 contains some examples associated with resilient design of a new system or facility, as opposed to resilience assessment of an existing system or facility, which is the primary focus of this section.

3.4.2 Assessment Examples of a Single System or Facility

A hazard without a vulnerability does not represent a risk, and so, vulnerabilities are the best place to focus finite resources for maximizing the risk reduction potential. The vulnerabilities and risks depend on the nature of the facility (e.g., risk regarding interruption of operations or some other goal such as ability to prevent a highly catastrophic accident). With this in mind, the following two examples are discussed in this section: a large medical campus and a nuclear power plant.

In terms of continuity of operations, vulnerabilities are identified through an impact assessment. The impact assessment outlines the components of the overall operating model and identifies the vulnerabilities based on the pertinent shocks and stresses. The output is a model of the operation identifying key inputs, outcomes, and stakeholders. This allows the impact of a shock or stress to be identified through the whole system so that the most vulnerable elements can be identified and assessed in terms of the maximum tolerable period of disruption (MTPD). The MTPD then becomes a useful performance measure against which resilience strategies can be measured. The following discussion illustrates how an MTPD-based criterion can be used for identifying the most vulnerable elements for various shock/hazard scenarios and thus helps make decisions about prioritized upgrades for improving the resilience of an existing facility. The MTPD construct is applicable to many systems/facilities such as the medical campus discussed in the following. On the other hand, for situations where a disastrous outcome is to be avoided (such as prevention of a nuclear accident after a beyond-design-basis/extremely rare hazard event), the disruption period could correspond to the duration during which the facility needs to maintain safety while it remains in an *island mode* (e.g., without access to power and external communication).

3.4.2.1 Operational Resilience of a Medical City. This example covers resilience assessment of a major medical facility, a *medical city*. The facility is intended to provide a state-of-the-art health care, as well as cutting-edge research and education facilities, with minimal disruption under specified threats/hazards. A resilience framework was used to illustrate the development of a resilience strategy to withstand identified shocks and stresses with minimal disruption. The main risks (shocks) are pandemics, natural hazards, and power failures, which can be ranked according to their potential impact on the planned facility. The identified hazards can then be used to inform the operational impact, an exercise in which all of the components affecting the resilience of the master plan at a strategic level need to be mapped in a system resilience diagram. This diagram is then used to illustrate how disruption to one element of the facility can cause disturbance to connected or dependent parts. This in turn makes it possible to propose targeted implementation of mitigation measures, such as hardening of critical assets, multipoint distribution and back-up power generation, and so on, and to increase the resilience of the medical facility. In addition, contingency measures such as major incident plans and business continuity plans can be identified, and the emergency response plans can be fed into operational procedures for the facility. The details are discussed in the following.

The desire for increased resilience is central to the previously described undertaking. If disruption, stoppage, or failure within the medical city are to be managed, a robust risk management and contingency planning approach is required that systematically identifies all potential sources of disruption, assesses the risks, identifies areas of greatest vulnerability, and provides mitigation strategies, as well as contingencies. Failure in any of these respects could expose the hospital to unnecessary disruption and without the relevant capabilities necessary to respond.

There are a number of long-term factors that could change the base assumptions upon which the medical campus was founded (called stress factors). Demographic trends provide the basis for continued population growth, which affects the overall demand and type of demands placed on the health-care facilities. Climate change and sea level rise are likely to increase the frequency and severity of severe weather events (e.g., increasing flooding risks as well as prolonging periods of no precipitation and increasing the potential for water stress). Other key concerns in terms of stress factors include rapid growth of obesity and its associated diabetes treatment (including potential need for amputations), the perceived threat to long-term energy supplies, and finally, a growth in antimicrobial resistance (for which a solution is not found). These factors could result in increased morbidity and mortality within hospitals and pose great difficulties for providing surgical treatments.

The key factors that could provide a shock to the campus systems can be categorized as health hazards, major accidents or terrorism, utilities and infrastructure failure, information hazards, and natural hazards. Health hazards, including pandemics, epidemics, new and emerging diseases, zoonotic diseases, or the release of biological or chemical agents, such as legionella, asbestos, or dangerous contaminants from industrial accidents, unsurprisingly represent the greatest risk to a hospital. Health hazards such as a pandemic have a dual impact on hospitals because of increased demand for health care albeit with reduced staff availability because of illness. Reduced staff levels also impact other key infrastructure increasing the likelihood of utilities failures.

Major accidents or terrorism include a wide range of hazards including industrial accidents, explosions, contamination, firearms attacks, and major transport accidents, all resulting in a consistent response from the emergency and security service with the evacuation and cordoning of hazardous areas, casualty and fatality management, a multiagency operation to resolve the hazard and the potential for a significant period of investigation with the cordon remaining in place until evidence has been gathered. These incidents would all result in significant disruption and more likely a complete cessation of all activities within affected areas.

Utilities and infrastructure failures are the external factors supporting the business or operation. Utilities include power, water and sewage, fuel, telecommunication, the internet, mobile network, and broadband. Infrastructure is mainly associated with mobility, that is, highways, air transport, public transport (e.g., bus, rail, metro, tram, and taxi), cycle ways, and pedestrian ways. Supply chains can also be included in this category.

Information-related hazards are those factors that can impact information systems. They include intellectual property, cyberattack, malicious software, instant communication technology failure, and data protection.

A natural hazard might have a negative effect on people or the environment. Natural hazards can be grouped in many ways, see Figure 1-2 for an example. Two broad categories are used herein: (1) geophysical hazards encompassing geological and meteorological phenomena such as earthquake, coastal erosion, volcanic eruption, tornado, and drought and (2) biological hazards relating to a diverse array of diseases and infestations. Other natural hazards such as floods and wildfires can result from a combination of geological, hydrological, and climatic factors. The three key natural hazards identified during the assessment of the medical campus were surface water and ground water flooding, snow storm, and earthquake.

The initial impact assessment for the medical campus identified the following areas where no disruption could be tolerated:

- Emergency departments,
- Certain surgical departments,
- Trauma departments,
- Critical dare units (including postanesthesia and resuscitation units),
- Acute wards, and
- Communicable disease departments.

Key on-site utilities are as follows:

- Power,
- Potable water,
- Telecommunications including information and communications technology and beeper or pager systems,
- Piped gases,

- Back-up generators,
- Stormwater drainage systems,
- Portable life support equipment,
- Regional highways and local road networks serving the medical campus,
- Helipads,
- Control rooms and security, and
- Off-site telecommunication provider (particularly if tele-surgery is desired).

The following functions (or functionalities) are deemed critical:

- Trauma units,
- Cardiac units,
- Dialysis units,
- Medicine and telemedicine,
- Logistics centers and antivirus guard systems,
- Kitchens and catering (dependent on time of day),
- District cooling, and
- Basement ventilation.

The resilience framework provided by Field et al. (2016, 2017) can now be used for resilience assessment. It is broken down into three main areas: (1) measures to reduce the overall exposure, (2) measures that increase resistance where there is exposure, and (3) contingencies that allow an effective response in the case of adverse hazard impact. These three areas are broken down into further areas for consideration and against which resilience can be measured, as shown in Figure 3-3. Using the aforementioned framework, preventive, mitigative, and adaptive strategies were identified, as presented in Tables 3-2 and 3-3.

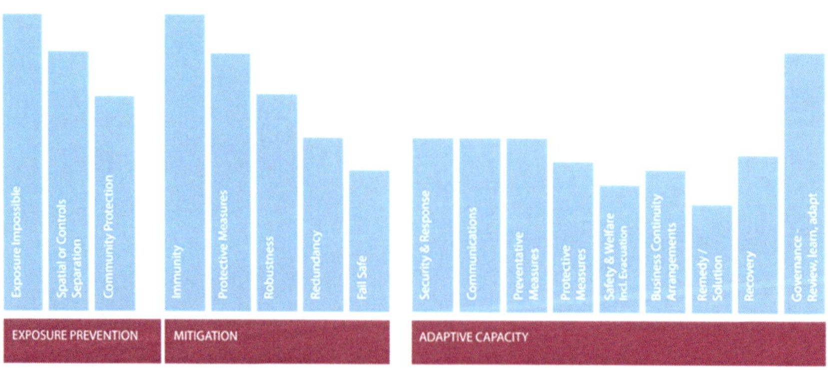

Figure 3-3. Resilience framework for assessing a hospital facility.

Table 3-2. Exposure Prevention and Mitigation Strategies for the *Medical City* Example.

Prevention/ Mitigation Strategy	Description
Diverse sources	Supplies and infrastructure from multiple sources to mitigate against the failure of a single organization
Diverse distribution	Supplies and infrastructure distributed through multiple routes to mitigate against the failure of a single route
Storage/spares in supply chain	Stores and spares are maintained as a buffer against a failure further up the supply chain and to ensure equipment failures within the supply chain can be quickly remedied
Regulations	Regulations to ensure the safe and secure operation of key infrastructure and supplies
Oversizing/spare capacity	Supply chains, utilities, and infrastructure are oversized or hold spare capacity to meet surge requirements and redundancy in case of partial failures
Maintenance capability	Suppliers and infrastructure organizations have the capability and capacity to ensure appropriate levels of maintenance
Protections	On-site measures and barriers
Maintenance—spares and capability	Capable of maintaining equipment and carrying sufficient spares to resolve issues within appropriate timescales. Where appropriate, procurement practices should take ease of maintenance into account
Security measures	Appropriate security measures and culture to deter criminal or malicious acts and prevent security breaches
Automatic measures	Measures that automatically deploy to provide a barrier, protecting vulnerabilities from immediate threats, for example, a fire wall against cyberattack or an automatic flood barrier

(Continued)

Table 3-2. (*Continued*) Exposure Prevention and Mitigation Strategies for the *Medical City* Example.

Prevention/ Mitigation Strategy	Description
Barriers	Permanent measures that provide a barrier, protecting vulnerabilities from threats by creating spatial separation
UPS	Uninterruptable power supply, ensuring continuous delivery of power for a limited time period to facilitate a controlled power down or continued operation when supported by back-up power generation
Generator capacity	Power-generating capability on-site to provide standby power in the event of a cut in power delivered to the site. Generators are not an instantaneous back up and if continuous power is required, they need to be used in concert with UPSs
Robustness	On-site robust design
High-tolerance networked utilities	Utilities networks designed with a high tolerance to shock factors, fluctuations in demand or construction/design errors
Diversification	Diversification strategies distribute critical elements to prevent single points of failure
Robust designs—dynamic modeling	Structural designs are modeled to prove their ability to withstand specific extreme shock factors such as wind, flooding, blast, heat, earthquake, and collision, where a threat has been identified
Environmentally resistant	Structural elements have been designed to be sufficiently durable to withstand the long-term effects of weather taking into account the changes in environment resulting from climate change within the design life of the structure
High-reliability equipment and infrastructure	Equipment and where appropriate, structural elements, are procured/installed to ensure strong reliability

(*Continued*)

Table 3-2. (*Continued*) Exposure Prevention and Mitigation Strategies for the *Medical City* Example.

Prevention/ Mitigation Strategy	Description
Redundancy	On-site redundancy
Storage (DOS, days of supply)	Stores maintained for the purposes of operational necessity or as contingency. These are quantified as the number of DOS the store represents against normal demand; high demand DOS can also be calculated separately
Looped networks	Utilities are designed in loops so that if there is a break or interruption at any point, supplies can be re-routed to ensure continuity
Multiple connection points and spare capacity	Spare capacity and multiple connection points ensure that if there is a failure at one connection point; full operational capability can be maintained through other connection points
Oversizing	Infrastructure capacity is oversized to ensure surges in demand do not cause failure. Oversizing is usually based on a reasonable worst-case risk assessment planning assumption. Structural elements can also be oversized to provide alternate load paths and prevent disproportional collapse
Expandable/ augmentable	Facilities can be expanded or augmented to meet surges in demand
Fail-safe	On-site fail-safe measures
Sumps and pumps	Sumps have been placed at low points where water could collect within infrastructure to allow collected water to be removed. Detectors can be used to warn when water is collecting and when sump capacity is reaching limits
Designed against progressive and or disproportionate collapse	The failure of a primary structural element does not cause disproportionate collapse of the wider structure

(*Continued*)

Table 3-2. (*Continued*) Exposure Prevention and Mitigation Strategies for the *Medical City* Example.

Prevention/ Mitigation Strategy	Description
Identified failure mode/sequence	Failure modes have been identified and understood so that appropriate mitigations have been designed to ensure cascading failure if prevented
Automatic shutoffs	Elements have been designed with automatic shutoffs to prevent further damage, harm, or loss if there is a failure
Safe close down	Systems are designed with automated safe close down procedures to prevent further damage, harm, or loss
Insurance	Plan/arrange for financial compensation in the event of a failure, disruption, or loss as a mitigation measure
Duplicate facilities/replication	A duplicate or replacement facility is available as a fallback (or can be build and operationalized in a very short amount of time). This could involve the rerolling of a lower priority facility
Standby	A fully functional facility is kept on standby

In addition, the following incident planning requirements were identified that need to be coordinated with the on-site major incident team:

- Multiagency command and control,
- Infection control,
- Evacuation,
- Mass casualties,
- Mass fatalities,
- Security, such as cordon procedures, and
- Highway network management.

The aforementioned approach enables determination whether the existing medical facility meets the selected MTPD threshold for a stipulated hazard scenario. Also, if the facility does not meet the MTPD requirement, then the approach helps identify the necessary preventive, mitigative, and adaptive measures (and their relative priorities in terms of risk reduction

Table 3-3. Adaptive Strategies for the *Medical City* Example.

Adaptive Strategy	Description
Control room/building management system (BMS)	Control room and BMS are in place to ensure failures and disruption are detected, located, identified, and monitored so that appropriate responses can be deployed
Command and control	Facilities and systems to ensure the overall management of an incident including situational awareness, control, coordination, communication, governance, and appropriate escalation
Contingency and recovery plans	Operational, tactical, and strategic plans to ensure continued operations despite disruptive factors and rapid recovery following an interruption, including tests, exercises, debriefs, learning, and adaptation
Evacuation	Coordinated evacuation plans at appropriate scales including floor, building, zone, and so on. Depending on scale, this may require specific assembly, marshaling, transport, and emergency shelter arrangements
Security and cordon protocols	Immediate action drills and standard operating procedures employed for any incident to ensure the safety and security of the public
Manual shutoffs/override	Systems and facilities designed with manual overrides or shutoffs if necessary, with appropriate security measure in place
Resolution capabilities	The capability to resolve the causes of damage, disruption, or harm so that the situation can be brought under control within recovery time objectives
Phased reduction in operations	Plans facilitate the phased reduction of operations ensuring controlled, proportionate responses

(*Continued*)

Table 3-3. (*Continued*) Adaptive Strategies for the *Medical City* Example.

Adaptive Strategy	Description
Surge demand phases and timescales	Plans facilitate the phased increase of operations ensuring controlled, proportionate responses
Public communication and sign posting	Warning and informing measures to ensure the public are aware of a hazard and its impacts
Invacuating and safe areas	Invacuating and safe areas/refuges may be a more appropriate solution to evacuation

potential) that need to be implemented to meet the MTPD threshold. It is noted that the same approach can be used for the design of a brand-new facility.

Another hospital resiliency assessment example using a rating system for seismic resilience is provided by Hassan and Mahmoud (2019).

3.4.2.2 Accident Resilience of Existing Nuclear Power Plants. The situation of nuclear power plants is a bit different in that, unlike the typical objective of minimal disruption to continued facility operation, here the overriding objective is to avoid a nuclear accident after an extremely rare beyond-design-basis event (i.e., in this case, rather than the usual power generation operation, the resilience in question has to do with plant's ability to maintain nuclear safety-related operations even if it is rendered into an island mode). This is typically achieved by maintaining the ability to keep the reactor in a safe shutdown mode (as well as maintain the spent fuel pool cooling capability) after the occurrence of a beyond-design-basis event. Loss of off-site power or cooling equipment poses major threats in terms of achieving these goals. As such, the nuclear power industry determined that an array of on-site defense-in-depth/adaptive features need to be in place for providing on-site power and shutdown cooling for a 72 h period after the reactor is shut down following a beyond-design-basis event. This stipulation is based on an expectation/experience that additional off-site resources can be mobilized within 72 h (e.g., ability of the local/federal authorities to provide further relief/back-up).

In the aftermath of the March 2011 Fukushima nuclear power plant disaster because of tsunami wave damage, the Nuclear Regulatory Commission (NRC) came up with a raft of policy initiatives and implementation procedures to ensure that the US fleet of nuclear power plants can avoid a similar fate in the event of various beyond-design-basis events (e.g., earthquake or flooding events that are more intense than the rare events that are already considered for the plant's design basis). With this in mind, the NRC introduced several requirements such as field walk downs to verify condition/accuracy of as-built plants, recalculation of extreme hazard probabilities to verify whether the existing plant continues to meet a minimum safety goal (e.g., a very low annual risk of core meltdown), implementation of targeted upgrades to achieve the minimum safety goal in the event the updated hazard probabilities indicate unacceptable safety risk, implementation of site-specific measures that reduce the risks associated with beyond-design-basis hazards, and so on.

In response to the requirement concerning diverse and flexible mitigation strategies, the Nuclear Energy Institute (NEI 2016) introduced an implementation guide (NEI 12-06) (NEI 2016) for the so-called FLEX methodology that helps ensure that multiple on-site sources for back-up power and cooling capability (e.g., cooling water, pumps, and so on) are available at each site, so that the plant safety is maintained for at least 72 h. An illustration of the FLEX methodology is shown in Figure 3-4 (NEII 2012). Among other things, this has meant that the existing plants needed to be retrofitted with storage facilities for back-up equipment that can be mobilized immediately after reactor shutdown in a beyond-design-basis event. A series of new administrative requirements and more enhanced

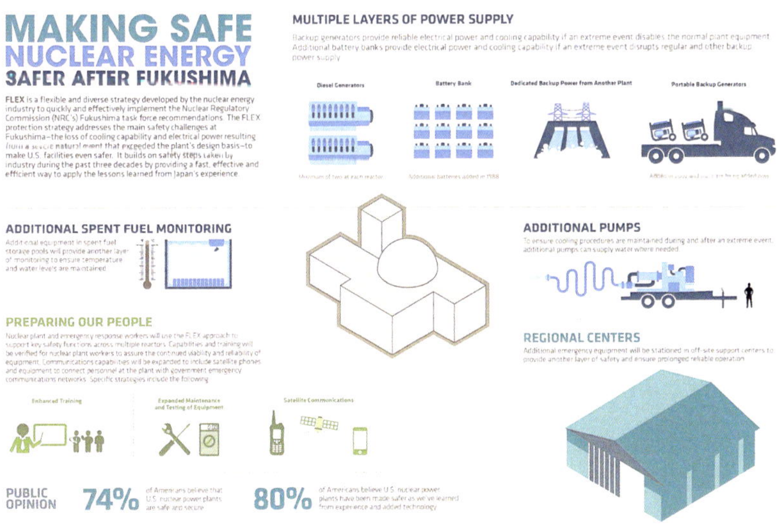

Figure 3-4. FLEX methodology for enhancing safety resilience of nuclear power plants.
Source: NEI (2016).

security requirements have also been implemented as a result (including those arising from new NRC requirements).

3.5 SYSTEM RESILIENCE ASSESSMENT METHODS

3.5.1 Analytical Considerations

Historically, the US government policy toward critical infrastructure security has focused on prevention and protection. However, following the terrorist attacks of September 11, 2001, the devastation from Hurricane Katrina in 2005, and a series of other disasters in the early 2000s, the infrastructure security community in the United States and globally recognized the impossibility of preventing all threats to all assets at all times. Consequently, critical infrastructure resilience emerged as a complementary goal to prevention-focused activities. Although critical infrastructure security activities primarily focus on preventing terrorism, accidents, and other disruptions, critical infrastructure resilience activities emphasize the infrastructure's ability to continue providing goods and services even in the event of disruptions. Together, critical infrastructure security and resilience strategies provide a more comprehensive set of activities for ensuring that critical infrastructure systems are prepared to operate in an uncertain, multihazard environment.

The increased emphasis on resilience has caused infrastructure stakeholders to ask frequently the following questions:

- How resilient is my infrastructure system?
- To which hazards is my infrastructure system most and least resilient?
- What can be done today to make my infrastructure system more resilient against the hazards of tomorrow?
- How should I prioritize infrastructure planning and design decisions to maximize infrastructure resilience, given limited time, budgets, and other constraints?

These questions and the prioritization of resilience in governmental and policy organizations have spurred infrastructure resilience research, and much of the research has focused on resilience metrics and assessment methodologies. Numerous methods have been proposed for this purpose, and this section describes the Infrastructure Resilience Analysis Method (IRAM) and how it has been applied to a variety of infrastructure systems and hazards. This section starts with briefly describing two broad categories of infrastructure resilience assessment models and identifies appropriate ones. Then, the section introduces the IRAM and details the definitions, metrics, and assessment processes included in the IRAM and ends with describing a case study in which the IRAM is applied to a freight rail network. This section offers conclusions potential extensions and research areas for the IRAM.

3.5.2 Infrastructure Resilience Analysis Method

This attribute- and performance-based method addresses the questions of "How resilient is my system?" and "What can I do to make my system more resilient?" (Vugrin et al. 2011, Biringer et al. 2013). The IRAM consists of four primary components:

- Resilience definition that highlights key features suitable for its quantification;
- Quantitative metrics for measuring resilience with system performance measures;
- Categories of infrastructure attributes that can help explain quantitative measurements and inform design efforts; and
- Assessment processes that formalize the application of the IRAM.

These components and other relevant concepts are described under separate headings.

3.5.2.1 Working Definition of Resilience. According to the IRAM, a working definition of resilience is a key starting point (Vugrin et al. 2011, Biringer et al. 2013): Given the occurrence of a particular disruptive event (or set of events), the resilience of a system to that event (or events) is the ability to reduce efficiently both the magnitude and the duration of the deviation from targeted system performance levels. This definition is in agreement with the PPD-21 (PPD 2013) definition. The IRAM emphasizes several features that ought to be captured in metrics

- Resilience is a contextual concept. Infrastructure resilience levels can and often will vary according to the system and the hazards being considered.
- System performance is a key factor in evaluating resilience. When analyzing infrastructures, system performance can be measured according to the goods and services that the infrastructure produces and/or delivers to consumers. Systems often adjust and reconfigure to cope with a disruption, so maintaining system structure is not as important as maintaining system performance. Hence, measurement of resilience includes evaluation of how a disruption affects system performance and causes infrastructure productivity to decrease relative to targeted system performance levels.
- Efficiency with which an infrastructure uses resources during disruption response and recovery activities affects the infrastructure's resilience; depending on the domain, these resources can be measured

in dollars, repair labor-hours, equipment, time, or similar metrics. If a disruption affects multiple infrastructure elements, resources may not be available to repair all impacted components simultaneously and resources may be scarce. All things being equal, a system that recovers more efficiently than another should be considered more resilient.

3.5.2.2 Quantification and Metrics. IRAM includes two sets of metrics for quantifying infrastructure resilience: systemic impact (SI) and total recovery effort (TRE) as schematically illustrated in Figure 3-5. For a specified disruptive event, the SI metric represents the cumulative impact of the disruption on the infrastructure's ability to provide goods and services. SI, defined in Equation (3-14), is measured by integrating the difference between a targeted system performance level (TSP) and the actual system performance level (SP) over the time period spanned by the onset of the disruption (t_0) and the completion of recovery activities (t_f). Oftentimes, infrastructure performance is measured according to multiple quantities, for example, a supply chain may produce multiple goods, power grids may deliver power to different classes of customers (residential, industrial, and critical), and so on. For these instances, the index j in Equation (3-14) is used to denote multiple performance measures, and q_j are weights that represent the relative importance of the performance measures, that is, larger q_j indicates a greater value or importance. If the system performance levels are measured in different units, the weights can also act as unit conversion factors:

$$\mathrm{SI} = \sum_j \int_{t_0}^{t_f} q_j(t)\left[\mathrm{TSP}_j(t) - \mathrm{SP}_j(t)\right]dt \tag{3-14}$$

The TRE metric represents the cumulative resources used by the infrastructure system to overcome and recover from the disruptive event. TRE, defined in Equation (3-15), is measured by integrating the recovery effort (RE) between t_0 and t_f. Similar to the calculation of SI, multiple recovery efforts, indexed by k in Equation (3-15), can be considered, and r_k

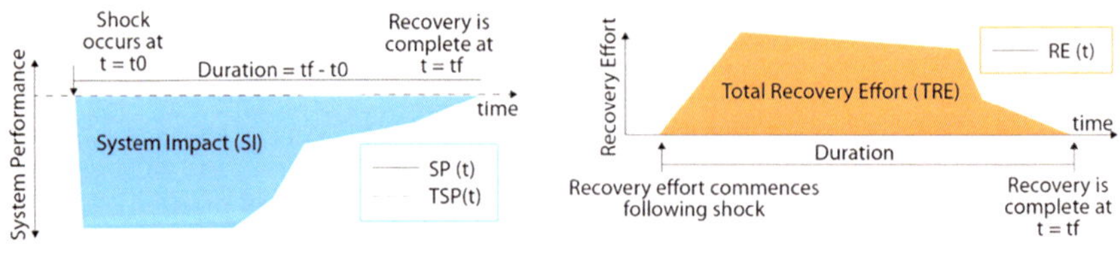

Figure 3-5. Systemic impact and total recovery effort metrics.

are weights representing the relative costs of the recovery measures. If various recovery efforts are measured in different units, the weights can also act as unit conversion factors:

$$\text{TRE} = \sum_{j} \int_{t_0}^{t_f} r_k(t)\left[\text{RE}_k(t)\right]dt \qquad \text{(3-15)}$$

The limits of the integrals in Equations (3-14) and (3-15) are the same because recovery effort could be started immediately after disruption. The interpretation of the SI and TRE quantities is fairly straightforward. Smaller SI values indicate that the infrastructure is better able to withstand and overcome the effects of the disruption; larger SI values indicate that the disruption has a larger impact on system performance. Smaller TRE values indicate that the infrastructure requires fewer resources and is more efficiently able to overcome the effects of the hazard; larger TRE values indicate that the infrastructure has to work harder to overcome the hazard.

SI and TRE provide different measures of the effect of a disruption on the infrastructure and can be considered separately. However, one may want to consider them together to provide a description of the total impact of the hazard on the system. Equation (3-16) combines quantities into a single resilience measure, R, by weighing SI and TRE with factor α and normalizing the results using a scaling factor N, as provided in Equation (3-17). The scaling factor enables comparison of different-sized systems and can be ignored when only considering a single system. A smaller α value indicates that system performance is more important than the usage of response resources; a larger α value indicates that response resources are limited and of greater importance than system performance. The structure of R is well suited for informing trade-offs between performance and response with associated efforts and costs

$$R = \frac{\text{SI} + \alpha(\text{TRE})}{N} \qquad \text{(3-16)}$$

$$N = \sum_{j} \int_{t_0}^{t_f} q_j(t)\left[\text{TSP}_j(t)\right]dt \qquad \text{(3-17)}$$

The following considerations should be noted when using these metrics to study infrastructure resilience:

- R and resilience of the system are inversely related. That is, larger R values imply lower levels of resilience and vice versa.

- Measurement of resilience in this manner implies that resilience is a continuous, not binary, concept. The metrics alone do not imply when a system is *resilient* or *not resilient*. Rather, these metrics are most informative when being used for comparative analyses, that is, informing which infrastructure design options are more/less resilient, does an infrastructure system exceed minimum resilience thresholds, against which hazards is the system more/less resilient, and so on.
- SI and TRE measure different quantities, but they are implicitly dependent on each other. Different REs result in different SP curves. [Note the functional dependence of SP, SI, and TRE on RE in Equations (3-14) and (3-15).] Thus, SI measures impact a specific RE. Hence, one ought to consider both metrics jointly.
- Oftentimes, the analyst may not be able to represent SP, TSP, and RE as mathematical functions but instead has time-series data to represent these quantities. In these instances, the integrals in Equations (3-14) to (3-17) can be replaced with summations for the same result.

The primary limitation of this method is using ad hoc mathematics not rooted in a mathematical theory, such as measure or probability theory. It should be used for simple system treatments that do not require using economic valuation and a formal benefit–cost analysis.

3.5.2.3 Resilience Capacities. The IRAM describes a set of four resilience capacities that can be used to understand and explain the results of quantitative analyses. The capacities are categories of infrastructure attributes that determine the infrastructure's overall resilience in one of the four ways.

An infrastructure's anticipative capacity consists of features that enable identification, prediction, and advanced warning of disruption conditions and enable a rapid, proactive response to hazards. Monitoring systems, threat intelligence, and forecasting are examples of system features that can strengthen anticipative capacities. The absorptive capacity consists of infrastructure attributes that help the system withstand or absorb the effects of a hazard. These attributes consist of relatively low effort options, such as redundancy, segregation, and excess capacity, which represent the preferred, go-to options. The adaptive capacity includes system properties that enable the system to reorganize and change the way it operates to overcome the effects of the disruption. Substitution, rerouting, conservation, and rationing are some examples of adaptive, resilience-enhancing features. The restorative capacity is the fourth capacity and includes system properties that facilitate system repair and recovery. Examples of restorative resilience-enhancing features include prepositioning, graceful degradation/shutdown, and reciprocal aid agreements. Table 3-4 lists the resilience capacities and examples of system features that affect the capacities.

Analysis of an infrastructure's resilience capacities can provide an understanding of how salient infrastructure features constrain and/or enhance the infrastructure's resilience. In doing so, analysts can develop options for enhancing resilience of the infrastructure. Note that the goal of resilience enhancement should not necessarily be maximizing each resilience capacity. Instead, analysts should aim to find the right combination of resilience enhancement features, given the infrastructure

Table 3-4. Resilience Capacities.

Resilience Attribute	Anticipative Capacity	Absorptive Capacity	Adaptive Capacity	Restorative Capacity
Capacity description	Ability to rapidly sense and identify hazards	Design features requiring few/no resources to withstand and absorb hazard effects	Ability to reorganize and modify operations to limit hazard effects	Ability to rapidly repair and reconstitute the system
Order of reliance	Before/during hazard manifests	First line of defense	Second line of defense	Third line of defense
Effort required	Varies	Automatic/little effort	Increased effort	Greatest effort
Example system features	Surveillance and monitoring systems; intrusion detection systems	Redundancy; decentralization; hardening/robustness; excess capacity; stored inventory	Substitution; rerouting; finding new suppliers; conservation or rationing; reorganization; ingenuity	Automated, self-repair systems; fail-safe modes; rapid forensics; graceful degradation; prepositioning; reciprocal aid

Source: Adapted from Biringer et al. (2013).

under consideration, the hazards that the analyst is considering, and the resources that are available to the infrastructure operators.

3.5.2.4 Assessment Process. The final element of the IRAM is the process with which one applies the IRAM. The process consists of seven primary steps as provided in Figure 3-6 as follows:

1. Specify analysis objectives: Resilience analyses can take on many forms, so the first step is specifying the questions the analysis aims to address. Examples of common infrastructure resilience analysis objectives include
   a. Measuring the resilience of a specified infrastructure system,
   b. Approximately comparing the costs and benefits of proposed disruption response strategies, and
   c. Ranking and/or optimizing infrastructure investment strategies to maximize resilience.

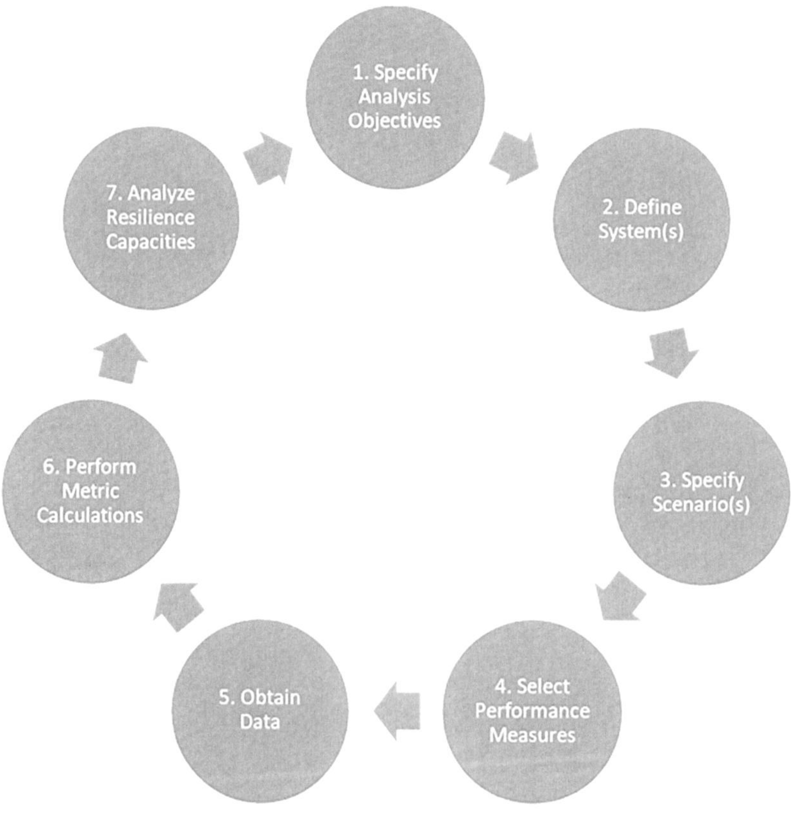

Figure 3-6. Systemic impact and total recovery effort metrics.

2. Define system(s): In the second step, the analyst specifies the infrastructure system under study and which pieces of the system are in and out of scope. Commonly, the analysts need to answer questions such as "What are the components of the infrastructure to be included?," "How are the components related and what are their (inter)dependencies?," and "What are the boundaries of the system?."
3. Specify disruption scenario(s): In the third step, the analyst specifies the disruption events to be included in the analysis. When specifying the scenarios, the analyst will need to answer "Which disruptions should be included?," "How do the disruptions directly and indirectly (through cascading effects) affect elements of the infrastructure system?," "What are the extents of the effects, and are they physical, functional, or virtual?," "How long do the effects last?," and "How does the infrastructure respond to the disruption?" Analysis of all postulated scenarios may not be possible, so scenarios must often be prioritized according to potential consequences, likelihood, and other factors.
4. Select performance measures: In this step, the analyst selects the performance measures [SP and RE in Equations (3-14) and (3-15)] and weighting factors [q, r, and α in Equations (3-14) and (3-16)] that are included in the IRAM metrics. The analyst also needs to specify targeted performances levels [TSP in Equation (3-14)].
5. Obtain data: The fifth step is the collection of system performance and recovery data for metric calculations. Data can be obtained from modeling and simulation platforms, physical test beds, historical data, and even expert judgment. When selecting the data source, the analysts should balance time, budget, and capability constraints with data quality. The results of the analysis are as good as the data being used.
6. Perform metric calculations: In the sixth step, the analyst calculates the resilience metrics according to Equations (3-14) to (3-17).
7. Analyze resilience capacities: In the final step, the analyst reviews infrastructure system attributes and identifies resilience enhancement features that affect the resilience of a system and lead to the quantitative results. Identification of these features provides guidance on how a system can be improved to become more resilient. This step may also identify behaviors of a system that were not considered previously (especially identification of recovery efforts) in the resilience analysis and may lead back to previous steps.

3.5.3 Case Study: Freight Railroads

This case study illustrates the IRAM using US freight railroads. It summarizes, modifies, and expands upon studies originally described in

Biringer et al. (2013) and Sandia Lab Technical Report SAND2010-6237 (Vugrin et al. 2010) by updating the resilience metrics considering an additional recovery strategy and discussing resilience capacities.

This case study focuses on the US freight railroads that transport goods across the United States, into Mexico and Canada, and to shipping ports and other transportation mechanisms for reaching other countries, playing a vital role in the United States and global economies. The sheer geographic expansive layout of the railroads exposes such a system to a variety of hazards and potential disruptions. The focus herein is on a hypothetical flooding scenario that damages four bridges on the Mississippi River. This scenario is of particular interest because the bridges included in the study are located near Chicago, the largest interchange point for east–west traffic. This case study aims to rank and identify preferred recovery strategies for repairing the bridges and restoring rail transports to their full capacity. The study uses a numerical optimization model to simulate the disruption of the network, repair of bridges and recovery of the network, and commodity flows across the network in the various states of network damage and repair.

3.5.3.1 State Analysis Objectives and Define System. The purpose of this analysis is to (1) quantify the impacts of the flooding scenario and bridge outages on commodity flows across the network, (2) identify the optimal strategy for repairing the bridges that maximize resilience of the network and compare, and (3) compare the optimal strategy to approaches where the railroads recover independently.

The system of interest, the US Class I freight railroads, is provided in Figure 3-7. By definition, Class I railroads had 2009 revenues exceeding $379M (AAR 2010). These railroads consist of more than 95,000 miles of track and are responsible for transporting the vast majority of railcar shipments in the United States (AAR 2018). The system includes tracks, rail yards, bridges, and tunnels that connect commodity origin locations to delivery destinations. Each of these physical elements have capacity constraints that limit the quantity of shipments at a given location and the time it takes to transport commodities from one location to another. The region of particular interest for this study is the network components and commodity movements in Nebraska, Kansas, Missouri, Illinois, and Iowa.

3.5.3.2 Specify the Scenario. This study assumes that four rail bridges on the northern Mississippi River are simultaneously damaged by flooding on the river. These bridges include the Union Pacific crossing at Clinton, Iowa; two BNSF bridges at Burlington, Iowa, and Ft. Madison, Iowa; and the Norfolk Southern bridge at Hannibal, Missouri. The bridges are assumed to be nonoperational, and no traffic can cross the bridges before they are repaired.

Figure 3-7. Freight rail network with the red rectangle denoting the region containing four bridges damaged because of flooding.

While the bridges are being repaired, the railroads will reroute shipments around the damaged bridges. This rerouting increases the distance that shipments must travel and also the in-transit time, resulting in increased operational costs. In addition, despite the rerouting adjustments by the railroads, the network may have insufficient capacity to transport all of the shipments. Hence, some railcars may not be able to be shipped, and the railroads will lose revenue for these cars.

This study assumes that the bridges can be repaired in one of the following three modes:

Normal Mode: This mode requires one repair crew and 15 days to complete. No traffic can cross the bridge during the 15-day period; after the repairs are complete, bridge capacity is restored to 100%. The cost of the repair is about $5M per bridge.

Emergency Mode: This mode requires two repair crews, but the repairs are completed within 10 days. No traffic can cross the bridge during the 10-day period, and bridge capacity is restored to 100% after the repair period. The cost of the repair is $10M per bridge.

Staged: This mode repairs the bridge in two stages. The first stage requires 1 crew and 9 days. After the first stage is complete, the capacity is restored to 50%. The second stage also requires a single crew and another 9 days. After the second stage, the capacity is restored to 100%. The repair cost for each stage is $3M, resulting in a total repair cost of $6M per bridge.

The case study assumes that each railroad company has one repair crew available.

This study considers three recovery strategies

1. Strategy 1: Each railroad repairs its bridges independently in the normal mode; all traffic are diverted around all bridges until they are all completely repaired.
2. Strategy 2: Each railroad repairs its bridges independently in the normal mode; traffic resumes across each bridge as it is repaired, regardless of the repair state of other bridges.
3. Strategy 3: The railroads cooperate and pool their resources to repair all of the bridges; traffic resumes across each bridge as it is repaired, regardless of the repair state of other bridges.

3.5.3.3 Select Performance Measures. For this study, system performance is measured as the total daily value of rail shipments through the Nebraska, Kansas, Missouri, Illinois, and Iowa. If the average revenue per railcar is estimated at \$1,770 per car (in 2011 US Dollars) this study defines SP(t) as number of railcars moved across the region on day t. The weight q in Equation (3-1) is set to \$1,770/car. TSP(t) as the average daily number of railcars moved across the region when the bridges are fully operational.

Hence, SI measures the total value of railcars not shipped during the recovery period.

Recovery for the network consists of bridge repair and rerouting. TRE measures the costs associated with these activities: (1) Bridge repair costs, $RE_1(t)$, depend on the repair mode selected by railroads. Normal and staged modes cost \$333 k/day/bridge, and emergency mode costs \$1 M/day/bridge. (2) Rerouting increases the distances traveled and in-transit times for shipments. Costs for rerouting can be calculated as follows: Increased travel distance costs $= q_2 RE_2(t)$, where $RE_2(t)$ is the additional car-miles traveled on day t and $q_2 = \$1.5$ per car-mile. Additional car-miles are measured relative to car-miles traveled on an average day in the five-state region when all bridges are operational. Increased in-transit time costs: $q_3 RE_3(t)$, where $RE_3(t)$ is the additional in-transit car-hours on day t and $q_3 = \$1.58$ per car-hour. Additional in-transit car-hours are measured relative to total car-hours for an average day in the five-state region when all bridges are operational.

Hence, TRE measures the total recovery costs for the railroad companies. The weighting factor α in Equation (3-16) is set to 1.

3.5.3.4 Obtain Data. Sandia Lab Technical Report SAND2010-6237 (Vugrin et al. 2010) used the Rail Network Analysis Tool (R-NAS) for simulating rail network flow disruption and restoration. R-NAS was also used to identify the optimal recovery strategy that maximizes resilience,

Table 3-5. Rail Transport Statistics for Undisrupted and Fully Disrupted Conditions.

Case	All bridges Fully Operational	Number of Bridges Operational
Car-miles traveled per day (1,000s)	17,300	17,820
Car-hours traveled per day (1,000s)	2,660	3,080
Cars not shipped	0	5,600

that is, minimizes Equation (3-16). For additional details on the R-NAS model, see Sandia Lab Technical Report SAND2010-6237 by Vugrin et al. (2010) and Jones et al. (2003).

Table 3-5 shows the average daily car-miles traveled, in-transit car-hours, and cars not shipped when the four bridges are fully operational and nonoperational. Daily car-miles traveled increase by approximately 3% when all of the bridges are damaged, and daily car-hours in transit increase by almost 16%. A total of 5,600 railcars are unable to be shipped each day when all bridges are nonoperational.

Figure 3-8 illustrates the bridge repair strategies for the independent and optimal, cooperative strategies. In the independent strategy, the railroads repair the bridges at Ft. Madison, Hannibal, and Clinton first using the normal repair mode. Because two BNSF bridges are damaged (Ft. Madison and Burlington), the Burlington Bridge cannot start to be

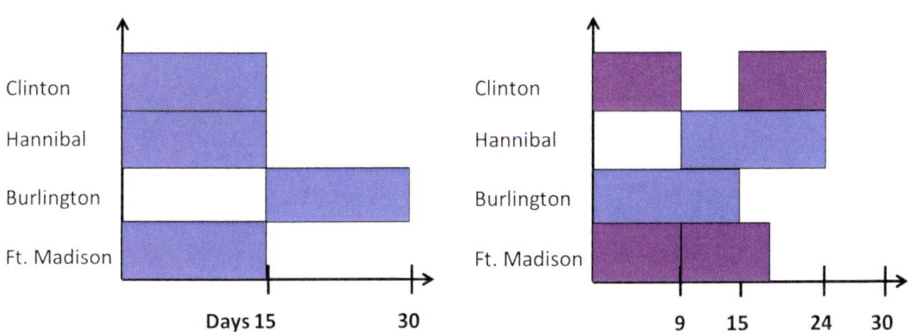

(a) Independent Repair (Strategies 1 and 2)

(b) Optimal, Cooperative Repair (Strategy 3)

Figure 3-8. Bridge repair strategies (blue for the normal repair mode and purple for the staged repair modes): (a) independent repair (Strategies 1 and 2), and (b) optimal, cooperative repair (Strategy 3).

Table 3-6. Impacts on Freight Rail Movements for Strategies 1 to 3.

Category	Strategy 1		Strategy 2		Strategy 3			
t (day)	0–15	16–30	0–15	15–30	0–9	9–15	15–18	18–24
TSP(*t*) – SP(*t*) (1,000s of cars/day)	5.6	5.6	5.6	0.98	5.6	0.55	0.08	0.08
$RE_1(t)$ ($M)	1	0.33	1	0.33	1	1	1	0.67
$RE_2(t)$ (additional 1,000 car-miles/day)	420	420	420	160	420	84	67	67
$RE_3(t)$ (additional 1,000 car-hours/day)	520	520	520	200	520	94	84	84

repaired until the Ft. Madison repairs are complete. The optimal repair strategy identified by R-NAS starts repairs on the Ft. Madison, Burlington, and Clinton bridges first. Ft. Madison and Clinton are repaired in the staged mode, so rail shipments can start moving across these bridges after just 9 days. Table 3-6 shows impacts of the bridge outages and repair strategies.

3.5.3.5 Quantify and Perform Metric Calculations. Table 3-7 shows the SI, TRE, and *R* results for Strategies 1, 2, and 3, respectively. These calculations use data from Table 3-6 and Equations (3-14) to (3-16). Strategy 1 has an SI value of almost $300M; this quantity represents the revenue lost by the railroads from not being able to ship railcars because of the damaged bridges. The TRE value, representing the cost of recovery, for Strategy 1 is more than $60M. The total impact of the bridge damage, *R*, is $360M for Strategy 1. For Strategy 2, independent repairs and immediate use of repaired bridges, SI, TRE, and *R* decrease to $176M, $48M, and $224M,

Table 3-7. Impacts on Freight Rail Movements for Strategies 1 to 3.

Strategy	Systemic Impact (SI)	Total Recovery Effort (TRE)	$R = SI + TRE$	Resilience Ranking
Strategy 1	$297M	$63M	$360M	3 (least resilient)
Strategy 2	$176M	$48M	$224M	2
Strategy 3	$96M	$43M	$139M	1 (most resilient)

respectively. Using the bridges as they are individually repaired decreases the number of railcars that cannot be shipped and decreases mileage and in-transit times that result from rerouting railcars, relative to Strategy 1. Under Strategy 3, SI, TRE, and *R* further decrease to $96M, $43M, and $139M, respectively. Using the staged repair mode decreases the number of railcars that cannot be shipped, in-transit hours, and mileage after 9 days, instead of 15 days as in Strategy 2. Furthermore, recovery is completed after 24 days under Strategy 3, 6 days sooner than under Strategies 1 and 2. Consequently, one can conclude that Strategy 3 is the most resilient repair strategy and that Strategy 1 is the least resilient repair strategy.

3.5.3.6 Analyze Resilience Capacities. Owing to space constraints, a complete analysis of the resilience capacities of railroads is not provided in this case study; instead, a few salient features of each capacity are highlighted as follows:

Absorptive Capacity: Redundant elements of the system have significant effects on the railroad resilience. The railroad system includes multiple tracks and bridges so that the loss of these four bridges does not completely halt traffic in the system.

Adaptive Capacity: Because redundancy exists in the system, the railroads can reroute traffic around the damaged bridges. Although rerouting comes with additional costs, these costs are smaller than revenues the railroads would lose by not shipping even larger numbers of railcars.

Restorative Capacity: Under normal conditions, antitrust regulations prevent cooperation across railroad companies. However, in times of crisis, railroad companies may provide assistance to other companies and share tracks. Strategy 3 demonstrates that cooperation and pooling of resources across railroads can enhance resilience to crisis events.

Anticipative Capacity: Although the case study, as described, did not delve into events prior to the onset of the flooding, weather forecasting and planning could bolster the system's anticipative capacity.

3.5.4 Observations and Limitations

Infrastructure resilience is a relatively new priority in the national security community, so resilience measurement and analysis belong to an active area of research and development. The IRAM has been developed to provide performance-based measures for quantifying resilience if infrastructure systems. Analysts can also use the IRAM's four resilience capacities to determine which infrastructure elements are constraining or enhancing the resilience of the infrastructure system. Although the IRAM is rooted in a formal mathematical theory for quantification, it offers simplicity and structure.

Although the case study focuses on railroads for identifying the most effective recovery strategies, in a resource-constrained environment, after the loss of four bridges because of flooding, the IRAM has been applied to many other infrastructure systems, including supply chains (Vugrin et al. 2011), hospitals (Vugrin et al. 2015), electrical power grids (Vugrin et al. 2014a), transportation systems (Vugrin et al. 2014b), and other infrastructure. Research opportunities exist to enhance the IRAM and other infrastructure resilience assessment methods, as follows:

- Extension to cyber and cyber–physical systems: The 2015 cyberattacks on the power grid in Ukraine and other infrastructure have highlighted the need to be able to analyze and quantify resilience of infrastructure systems to cyberattacks.
- Improved modeling of human behaviors: Human behaviors and response can significantly improve or worsen infrastructure operations during disruptions. Better representation of human behaviors during disruptions can improve our understanding of infrastructure resilience.
- Characterization of key uncertainties: Infrastructure performance for extreme events can be difficult to precisely characterize, so the including sources of uncertainty can help describe the range of possibilities.

3.6 SYSTEM-OF-SYSTEMS ASSESSMENT METHODS

This section introduces an SoS approach for assessing infrastructure resilience. Integration of methods from system engineering and system science provides a comprehensive lens through which complex characteristics and interrelationships among various systems and processes influence infrastructure resilience. To this end, this section starts by introducing fundamental definitions and attributes of system of systems. Then a taxonomy for SoS analysis of infrastructure resilience is discussed. Afterward, a framework explaining the steps for SoS analysis of infrastructure resilience is presented. Finally, some examples regarding the applications of SoS analysis of infrastructure resilience are presented.

3.6.1 Distinguishing Attributes of Systems of Systems

The first step in system analysis is to determine the type of the system (e.g., monolithic versus systems or system of systems (SoS) (Zhu and

Mostafavi 2018). Different types of systems have different traits. Maier (1998) characterized an SoS as a combination of a set of managerially and operationally independent systems, which performs a function not performed by a single system alone. Civil infrastructure is an SoS because of the existence and interaction of several independent/interdependent systems and human actors. The adoption of an SoS perspective facilitates a bottom-up assessment in which different dimensions of analysis pertaining to resilience-based engineering and planning of infrastructure could be investigated (Mostafavi et al. 2011). The traits of an SoS are presented in this section to determine whether infrastructure resilience can be characterized as an SoS problem. Rechtin (1991) defined a monolithic system as "a set of different elements so connected or related so as to perform a unique function not performable by the elements alone." Along the same lines, DeLaurentis and Callaway (2004) characterize an SoS as "the combination of a set of different systems [that] forms a larger system of systems that performs a function not performable by a single system alone." Maier (1998) defines an SoS as "an assemblage of components which individually may be regarded as systems, and which possesses two additional properties: operational independence of components ... and managerial independence of the components." SoSs have different traits than monolithic systems. Civil infrastructure is an SoS because of the existence and interaction of several independent or interdependent systems and human actors whose interactions affect resilience, as shown in Figure 3-9 (see also Figure 2-2). Figure 3-9 shows the complex relationships among various human actors and their decision-making processes in response to the evolving hazard landscape affecting the resilience performance of infrastructure systems. Hence, the creation of an SoS approach provides a better theoretical lens and facilitates resilience assessment in infrastructure systems (Mostafavi et al. 2011, 2012; Zhu and Mostafavi 2014).

Since the emergence of SoS, different categories of its distinguishing traits have been introduced. Maier (1998) cites geographical distribution, emergent behavior, evolutionary development, operational independence, and managerial independence as distinguishing characteristics. Boardman and Sauser (2006) cite autonomy, belonging, connectivity, diversity, and emergence as distinguishing characteristics that are generally the same as those introduced by Maier (1998). Maier (1998) also concluded that managerial independence and operational independence are primary and necessary conditions for an SoS setting; and emergence, evolutionary behavior, diversity in geographical distribution, and connectivity are secondary conditions. If a setting lacks the primary conditions, Maier contends, it cannot be considered an SoS even if it has the secondary characteristics. Maier (1998) defines operational independence as follows: "if the SoS is disassembled into its component systems, the components

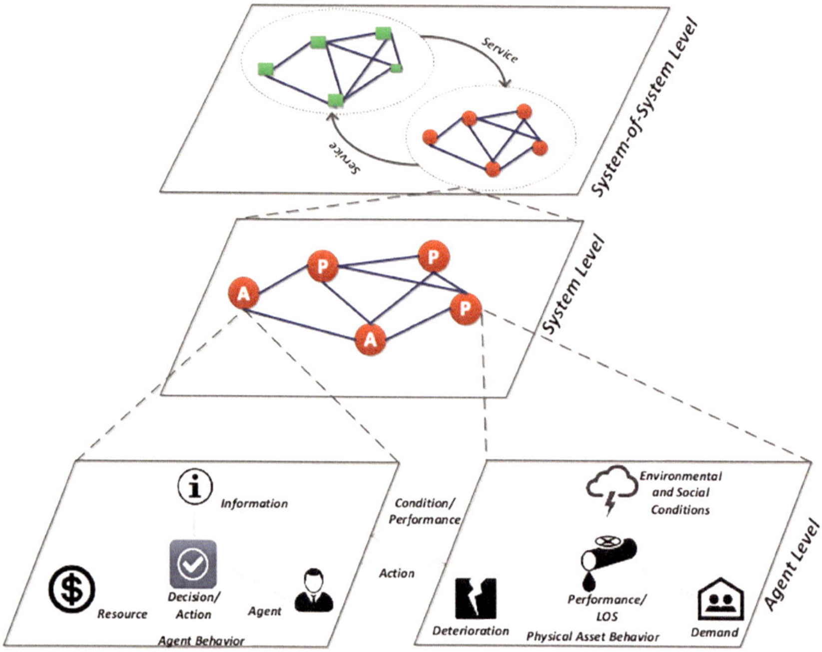

Figure 3-9. Conceptualization of SoS for resilience assessment of infrastructure.

systems must be able to usefully operate independently," and managerial independence as follows: "the component systems not only can operate independently, they do operate independently." An SoS performs functions and carries out purposes that do not reside in any component system. These behaviors are the emergent properties of the entire SoS and cannot be localized to any component system. The principal purposes of the SoS are fulfilled by these behaviors. SoSs can be characterized from another aspect as directed, collaborative, and virtual. Identifying whether an SoS is directed, collaborative, or virtual enables a better understanding of SoS dynamics and structure. A directed SoS has a centrally managed system to assure that the SoS function is performed, whereby the component systems functions are subordinated to the SoS function. A collaborative SoS lacks a centrally managed system. It is the collaboration between the key component systems and the human actors that provides an enforcement mechanism to fulfill the SoS function. Finally, in a virtual SoS, there is not a centrally agreed upon purpose for the SoS. In the context of civil infrastructure, at the local level and within a sector, systems can be considered as directed (a particular agency, such as transportation agency, manages an SoS of transportation systems). However, at regional and national levels and considering interdependencies among different sectors, systems would be categorized as a collaborative SoS, although the extent of collaboration and coordination among the actors who govern and manage the systems may not be extensive.

Given the SoS attributes discussed previously, it is evident that civil infrastructure has all the distinguishing attributes of SoS, and hence, an SoS perspective is an important step in resilience assessment. The next section presents a taxonomy for conceptualization of civil infrastructure resilience through the lens of an SoS perspective.

3.6.2 Taxonomy for Resilience Assessment

Three dimensions can be investigated in an SoS taxonomy for assessing resilience in infrastructure systems (Figure 3-10 or Figure 2-2). First, according to Figure 3-10, resilience analysis can be implemented with respect to different *classifications* in terms of infrastructure sectors, assets, performance conditions, and natural hazards. Thus, the context of the analysis for this dimension is defined to include the classifications of sectors (e.g., transportation, water, and power), assets (e.g., highways, roads, bridges, and water mains), phases of disasters (e.g., preparedness, mitigation, response, and recovery), and types of natural disasters (e.g., hurricane, flood, and earthquake). Resilience analysis could consider a single sector (road networks) or multiple sectors (interdependencies among flood control, transportation, and emergency facilities), or it may

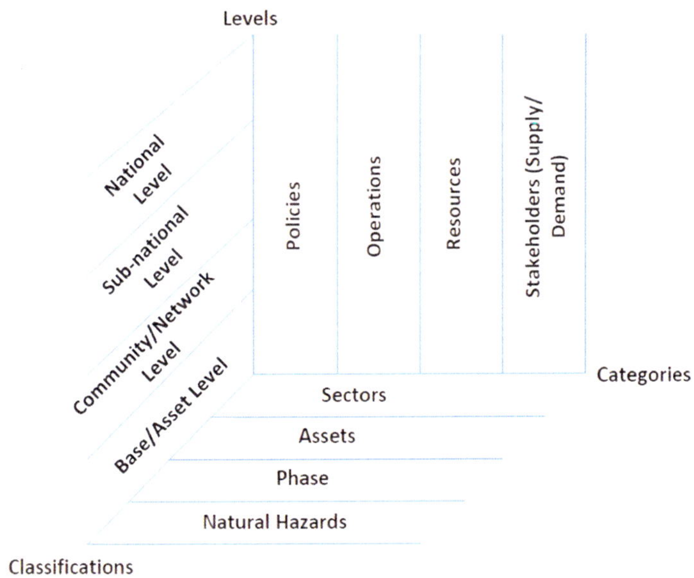

Figure 3-10. Dimensions for assessing the resilience of infrastructure.

investigate a single asset or multiple assets. Also, the analysis could consider the resilience of assets to one class or multiple classes of natural disasters. The greater the number of classifications considered, the greater the complexity of the analysis.

The second dimension of the SoS taxonomy includes different *categories* of components, and entities in the analysis are explored and abstracted (see Figure 3-10). The component categories include resources, stakeholders, operations, and policies that exist across hierarchical *levels*. *Resources* include the nonhuman entities that facilitate rehabilitation, retrofit, and capacity expansion actions and could be physical or financial. *Stakeholders* are human and organizational entities whose goals, decision making, actions, and interactions affect the planning and resource allocation of infrastructure. *Operations* include the application of the decisions or intents of *stakeholders* to direct the use of *resources*. *Policies* (plans and institutions) are external forces affecting the availability of resources and the decisions of stakeholders (DeLaurentis 2005, Mostafavi et al. 2016).

The third dimension pertains to the *levels* of analysis (see Figure 3-10), which includes the asset, network, subnational, and national levels. In the proposed SoS approach, the resilience at each level is obtained by aggregating the resilience at the level below it. This necessitates a definition of the resilience measures at each level. The multilevel aggregation of resilience in the proposed SoS approach enables the resilience of infrastructure systems to be integrated with the other dimensions of resilience (e.g., household and societal resilience) and combined into an SoS resilience assessment.

3.6.3 Method for System-of-Systems Resilience Assessment

Based on the taxonomy presented in the previous section, the resilience assessment method is based on a framework as discussed here to show the necessary steps. The SoS framework identifies different system elements and attributes that need to be abstracted and modeled to enable conducting exploratory analysis on civil infrastructure. The proposed SoS framework aims to provide a comprehensive approach for conducting infrastructure resilience analysis. The definition phase of the framework determines the specific measures and objectives that need to be taken into consideration for examining the resilience of infrastructure. The abstraction phase specifies the dynamic mechanisms and complex relationships among hazards/stressors, physical networks, and human decision makers. The implementation phase involves simulation modeling, as well as analysis steps based on robust decision-making (RDM) techniques to identify resilience pathways. Each phase includes a number of tasks described in detail in subsequent subsections.

3.6.3.1 Definition Phase. The first phase of the analysis is definition. The outcomes of the definition phase will inform the relevant stressors, actor and infrastructure attributes, and metrics to be considered in the abstraction and implementation phases. This phase includes two tasks: (1) defining the levels of analysis, such as base, system, and SoS, the context of analysis, and limitations and (2) defining the metrics for evaluating SoS performance and resilience at different levels of analysis (see Figure 3-9). For example, the attributes and interactions of institutional actors and physical infrastructure affect the resilience outcomes at the system level. The context of the analysis should define the infrastructure sector as well as the hazard's impacts for which the analysis is performed. The context of the analysis determines the type of stressors to be included in the analysis, the impact of stressors on physical infrastructure, and the action space of the institutional actors for responding to stressors. For example, assessment of water infrastructure systems under sea-level rise impacts would involve different stressors, physical infrastructure impacts, and action space compared to examining road networks performance under the impacts of temperature variation. The second task in the definition phase is to define the metrics for evaluating resilience and performance across different levels. Consideration of different resilience metrics at different levels would depend on the study objective and context. For example, Batouli and Mostafavi (2016) used a network-level life-cycle cost as a metric for evaluating the impacts of flooding on road infrastructure to determine the value of adaptation actions. Other studies (Dehghani et al. 2014) have used measures of network vulnerability for assessment infrastructure resilience under disruptions caused by natural disasters. Another important consideration is the relationship between different metrics at different levels. Because of the nonlinear behaviors in civil infrastructure systems (CIS), the resilience metrics at each level cannot simply be determined by aggregating the metrics at the levels below. In other words, resilience performance at the SoS level is an emergent property as a result of the interactions between different systems components at the level below. The aggregation of individual systems resilience may not be an indicator of CIS at the SoS level.

3.6.3.2 Abstraction Phase. The second phase of the SoS framework is abstraction. In the abstraction phase, relevant institutional actors and physical infrastructure assets and their attributes and interactions at the base level are captured. There are various attributes and behaviors that affect the internal feedback processes between institutional actors and physical infrastructure assets. For institutional actors, the decision-making behaviors such as information processing, resource allocation, project prioritization, and retrofit/capacity expansion are examples of behaviors that may be abstracted. An important aspect of SoS analysis of CIS resilience is the ability to integrate asset condition degradation, level of service (LoS), and vulnerability with the decision-making processes and adaptation actions of institutional actors and enable dynamic analysis over time (Koetse and Rietveld 2009, Lambert et al. 2013, Dehghani et al. 2014).

3.6.3.3 Implementation Phase. The third phase of the SoS framework is implementation in which computational representation of abstracted system components is created for conducting simulation experiments and exploratory analyses. An important step in the implementation phase is the selection of appropriate modeling and simulation methods. The selected modeling techniques should be consistent with the characteristics of the system. In the assessment of the impacts of hazards on infrastructure systems, an appropriate modeling technique should capture the dynamic, stochastic, and adaptive nature of system attributes. To this end, different modeling methods can be used for different system components and integrated into a multimethod model.

For modeling the performance of infrastructure assets, system dynamics, Markov chain, and mathematical modeling are examples of modeling techniques that can be used (Mostafavi and Ganapati 2019). For example, Rehan et al. (2011) and Rashedi and Hegazy (2016) utilized system dynamics for modeling the performance of water distribution infrastructure assets. Ortiz-García et al. (2006) used dynamic mathematical approaches to model the condition and deterioration of road pavements. For implementing the decision making and behaviors of institutional actors, agent-based modeling (ABM) can be adopted. ABM is an effective simulation approach for analyzing decision-making processes of actors in infrastructure systems (Pahl-Wostl 2002, Sanford Bernhardt and McNeil 2008, Batouli and Mostafavi 2014, Mostafavi et al. 2014, 2016, Bhamidipati et al. 2016). Using ABM enables (1) discovering what decision rules, microbehaviors, and preferences affecting adaptation decisions; and (2) juxtaposing the preferences of various decision makers with the range of mitigation and adaptation alternatives to determine the distribution of expected outcomes. The selection of an appropriate modeling approach for implementing each component is affected by the ability to integrate the modeling techniques into a multimethod simulation platform. A robust multimethod simulation platform should be able to cope with the complexity of calculating dynamic variables and uncertainties from different sources at different levels of multiple subsystems and modeling methods.

3.6.4 Model-Based Exploratory Analysis

Exploratory analysis has been developed in the literature related to robust decision-making (RDM). Exploratory analysis uses computational models and simulation experiments to conduct scenario analysis and evaluate the behavior of complex systems under uncertainty (Bankes 1993, Agusdinata 2008, Mostafavi et al. 2014). Exploratory analysis is a model-based method for decision support. It is an approach that utilizes different simulation methods (e.g., system dynamics and ABM) for decision making in problems that include deep uncertainty (Mostafavi and Ganapati 2019). Traditionally, simulation models have been used for predictive analysis; however, exploratory analysis focuses on more explorative use of models based on different future scenarios. Exploratory analysis is an important component for RDM under deep uncertainty (Lempert et al. 2004, Kwakkel and Pruyt 2013). The objective of RDM is to stress test the systems under different possibilities and to examine system features and relationships that led to desirable system outcomes under various scenarios. Hence, through model-based stress testing, SoS assessment enables examining thousands or even millions of scenarios composed of different hazards and system characteristics to evaluate strategies that lead to resilient outcomes under a majority of scenario-given constraints.

The components of the proposed SoS framework can be implemented through using multiple simulation and modeling methods. There is no generalized method or model that can capture all the attributes, complex behaviors, and dynamic interactions in civil infrastructure. Depending on the objectives of an exploratory analysis, as well as system elements and attributes considered, appropriate simulation models can be adopted and integrated. The ability to conduct exploratory analysis is the most important advantage of the proposed SoS framework. In SoS analysis of infrastructure system resilience, the results of simulation models should be processed to generate different possibilities and to identify the decision factors affecting resilience. To this end, exploratory analysis of infrastructure resilience explores the outputs of different scenarios by conducting thousands or millions of computational experiments that help in analyzing the system behavior. The process of exploratory analysis includes different steps (Figure 3-11). The data obtained from simulated data can be analyzed through various statistical approaches to conduct metamodeling. To this end, metamodeling of simulated data can provide insights into the significance of various elements affecting the resilience of infrastructure under disaster impacts. Metamodeling enables identifying robust pathways across multiple scenarios, assumptions that lead to a certain output, and key trade-offs across pathways (Rasoulkhani et al. 2018).

3.6.5 Examples of System-of-Systems Resilience Assessment

The adoption of SoS perspective in assessing infrastructure resilience has been increasing in recent years. Two example studies are presented here. In the first study, Batouli and Mostafavi (2018) adopted an SoS

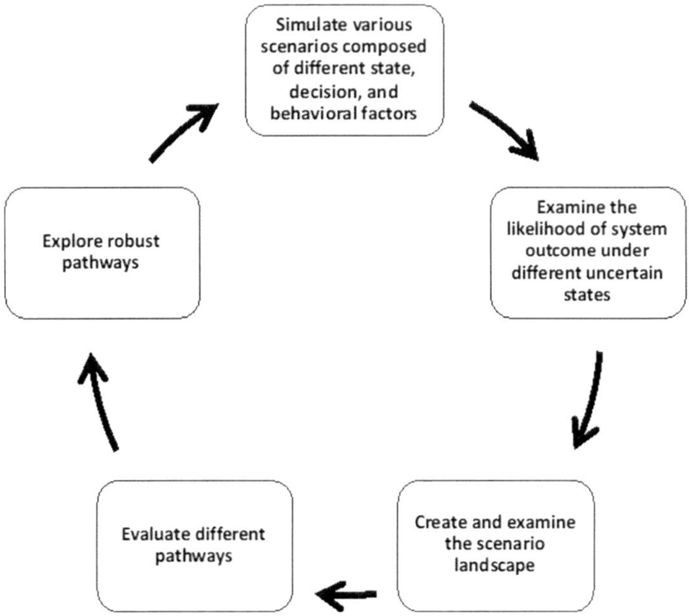

Figure 3-11. Exploratory analysis steps.

approach in assessing the resilience of a network to the long-term SLR impacts on coastal areas. In this study, the SoS was modeled using a multiagent simulation model, as shown in Figure 3-12. The simulation model was then used in exploratory analysis and stress testing of long-term life-cycle costs under different scenarios of sea-level rise and adaptation investments in a case study of the City of Miami Beach, as shown in Figure 3-13.

The second example study is related to evaluation of the long-term resilience of water distribution networks. Rasoulkhani and Mostafavi (2018) and Rasoulkhani et al. (2019) adopted an SoS approach and developed a multiagent simulation model (Figure 3-14) for evaluating the effects of population fluctuation and funding gaps on the long-term performance and resilience of dual water distribution networks of potable and nonpotable water with annual expenditure and annual capital expenditure (CapEx) (Figure 3-15).

These two examples demonstrate the adoption of an SoS assessment in single infrastructure sectors. However, the same SoS approach and model-based exploratory analysis can be adopted for analyzing resilience in multiple infrastructure at city and regional scales. In fact, the SoS approach enables model integration toward a convergent model of city and regional resilience.

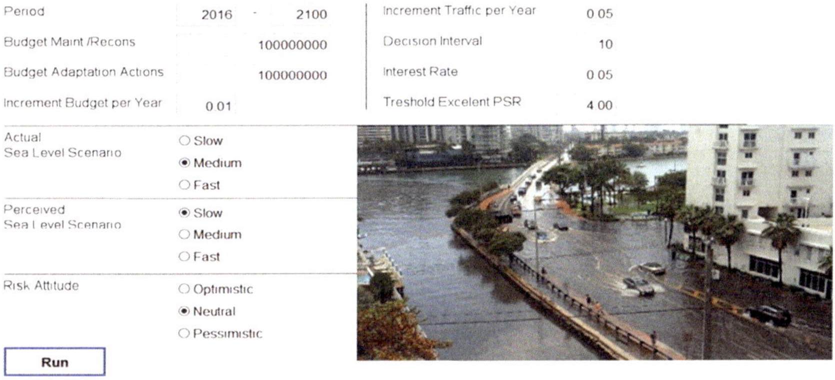

(a) Model Interface

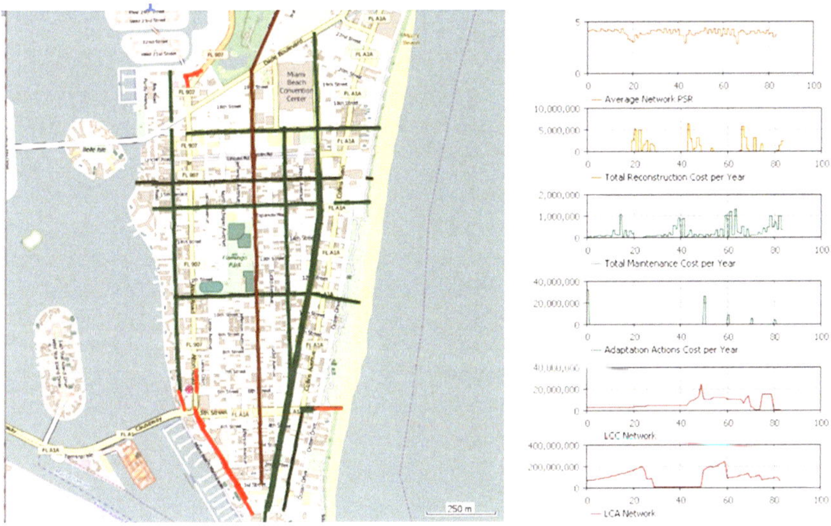

(b) Animation and Visualization

Figure 3-12. Simulation for assessing a road network resilience to SLR for 84 years: (a) model interface, and (b) animation and visualization.

3.7 INFRASTRUCTURE NETWORK TOPOLOGICAL VULNERABILITY AND RESILIENCE METHODS

3.7.1 Terminology

Key terminologies that are used in infrastructure network analyses are provided in this section. Key terms from network perspectives include network, efficiency, robustness, vulnerability, and resilience.

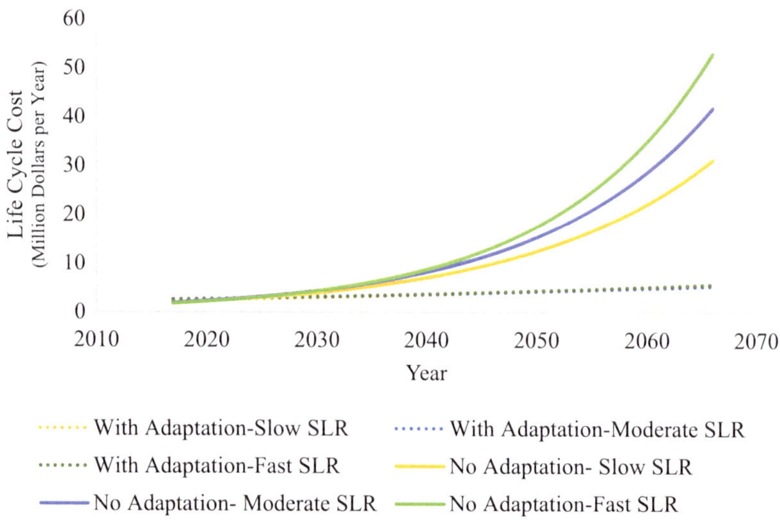

Figure 3-13. Exploratory analysis of life-cycle cost impacts of SLR and adaptation actions in the road network of Miami beach for 50 years.

A network or graph, in its simplest form, is a collection of points joined together in pairs by lines. Points are referred to as vertices or nodes, and lines are termed links or edges (Newman 2010). A network is a powerful abstract representation of a particular simplified system that breaks the system into nodes and links with the pattern of connections. The behavior

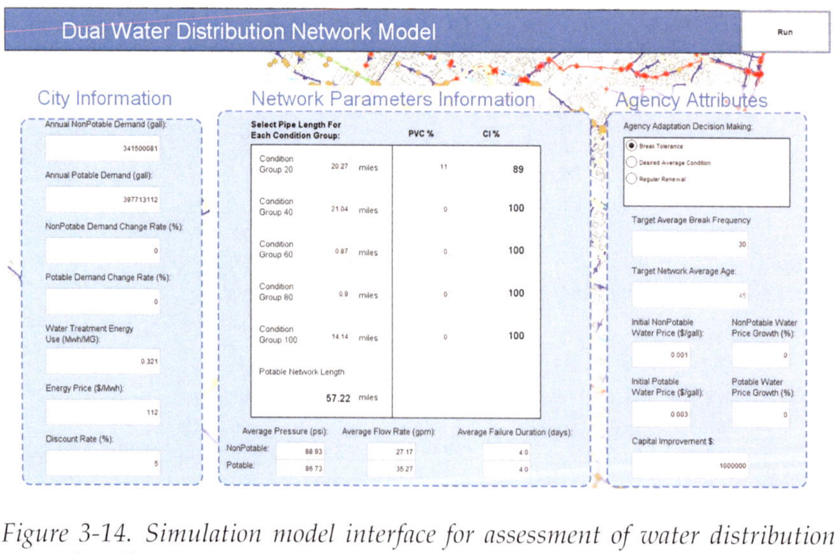

Figure 3-14. Simulation model interface for assessment of water distribution network resilience.

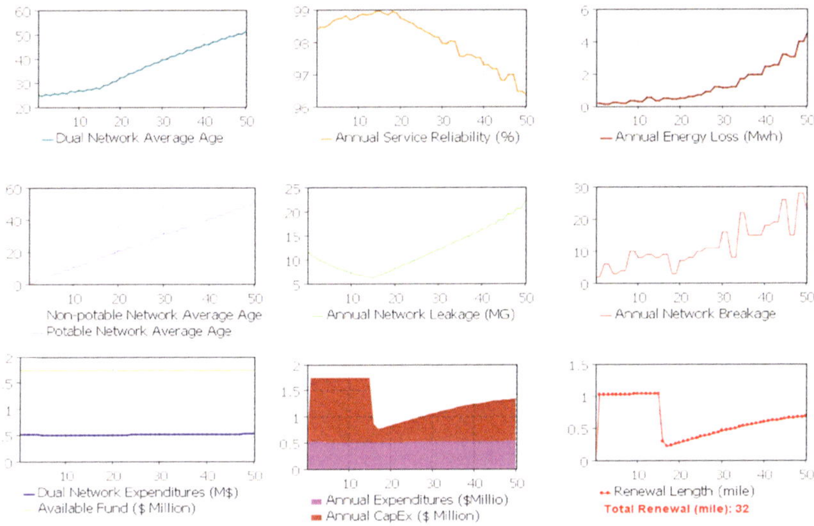

Figure 3-15. Model output interface for scenario analysis of performance measures of water distribution networks for 50 years.

of any given system is highly dependent on those patterns of connection and interaction that represent the most significant aspects of any specific network's structure. Thus, defining network properties and utilizing methods and tools to determine the network patterns of connection are necessary. Most infrastructure systems could be intrinsically representative of network paradigms because their features represent network components and attributes.

The network efficiency or global network efficiency (E_G) is a measure of how efficient communication or flow is between any two nodes in a network (Latora and Marchiori 2001).

Robustness is the ability of the system and system components to sustain external shocks without significant degradation of performance (Tierney and Bruneau 2007, Ayyub 2015). The robustness of a network, therefore, is its ability to maintain residual node connectivity after a possible adverse event resulting in disruptions. To enhance resilience, improving robustness is of paramount importance.

Vulnerability refers to the inability of a system to withstand the effects of an adverse event causing disruptions. Vulnerability is defined as an internal risk factor of system components that are exposed to external shocks and are more susceptible to be affected (Paul 2014). The concept of vulnerability for networks is increasingly being used to quantify the impact of any adverse event on the performance of networks. In an infrastructure network, vulnerability relates to the negative changes in the global connectivity of a network after perturbations. Compared to reliability analysis which is a probabilistic measure of the network connectivity, vulnerability analysis is more focused on the consequence of abnormal events instead (Jenelius et al. 2006, Jenelius and Mattsson 2015).

Resilience as a property of such networks is defined by PPD-21 in Chapter 2 (PPD 2013) as "the ability to prepare for and adapt to changing conditions and withstand and recover rapidly from disruptions." Network resilience has been defined in the literature (Gao et al. 2016, Ayyub 2015, Gilbert and Ayyub 2016). In a network, resilience refers to its ability to maintain sufficient level of network connectivity to sustain its vital functionality when a random or intentional disruption emerges.

3.7.2 Methods for Quantifying Network Resilience

This section provides methods for quantifying network topology resilience that are presented in terms of a general framework, as shown in Figure 3-16, and illustrated concurrently using a metro (i.e., subway) network. A systematic approach for assessing the resilience of a network requires characterizing its form, size, dependency, and interdependency

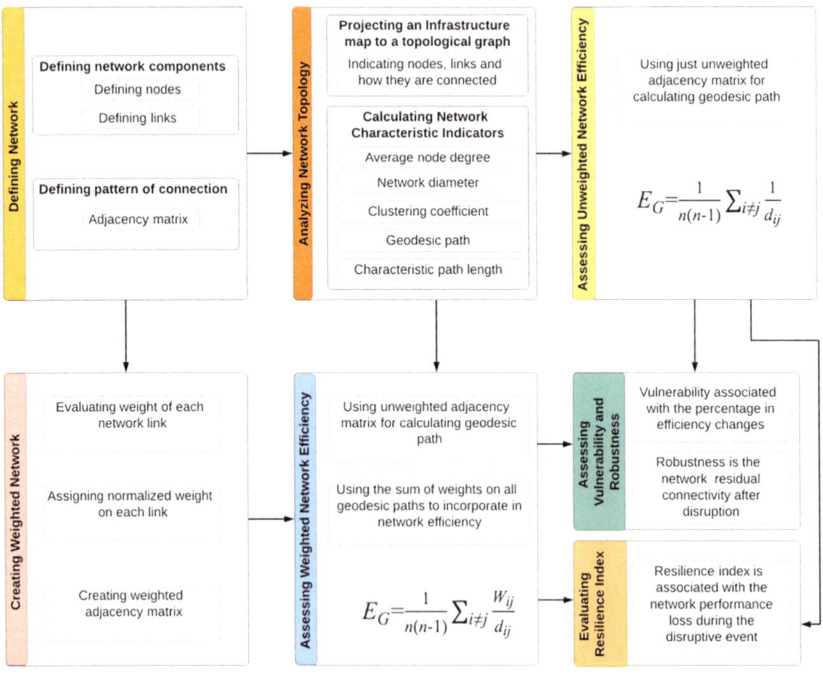

Figure 3-16. General framework and methodology of quantifying infrastructure network resilience.

under normative sources of disruption. The size and complexity of a system, along with a topological analysis, determine appropriate metrics for the resilience of a network (Saadat et al. 2019, 2020). The methodology proposed in this section is based on complex network theory (CNT) that represents a real system of interest in the form of a graph with nontrivial topological elements. The framework consists of the following steps:

1. Defining the infrastructure network as nodes and links representing pattern of connectivity among nodes by links;
2. Analyzing topology by computing node degree, average node degree, and adjacency matrix offering a basis for calculating characteristic path length and network efficiency;
3. Assessing the network efficiency of unweighted networks, which is required for calculating vulnerability;
4. Evaluating weight for each link of the metro network (as an example) using the data available on each station to generate the weighted adjacency matrix;
5. Assessing the network efficiency of weighted networks, which is the basis for measuring its vulnerability;
6. Assessing network vulnerability and robustness based on network efficiency changes prior to disruption and following disruption; and
7. Evaluating the resilience metric as an index for an infrastructure network based on vulnerability and robustness analysis.

Weighted networks are networks in which links carry some weights, and unweighted networks are free of this characteristic. For analyzing unweighted networks, Steps 4 and 5 are not required; and for analyzing weighted networks, Step 3 is unnecessary.

3.7.2.1 Defining Network Topology. Network components can be defined by vector G:

$$G = [S, E] \tag{3-18}$$

where S is the set of nodes and specified as

$$S = [s_i \mid i = 1, 2, 3, 4, \ldots, n] \tag{3-19}$$

and E is the collection of all links such as

$$E = [e_{ij} \mid i, j = 1, 2, 3, 4, \ldots, n] \tag{3-20}$$

where e_{ij} indicates the link that connects node i to node j.

The pattern of connection between any two nodes in a network is represented by the adjacency matrix. In complex network theory (CNT), an adjacency matrix is a square $n \times n$ matrix used to illustrate the state of connection in an entire network. For unweighted networks with n nodes, in which the links carry no weight, an adjacency matrix is denoted as $A_{ij} = [a_{ij}]_{n\times n}$, where

$$a_{ij} = \begin{cases} \infty & \text{for nodes } i \text{ and } j \text{ not connected directly} \\ 1 & \text{for a direct link between nodes } i \text{ and } j \\ 0 & \text{for } i = j\text{, connecting a node with itself} \end{cases} \quad (3\text{-}21)$$

On the contrary, a weighted network in which each link carries a specific weight is represented by a matrix element as follows:

$$a_{ij} = \begin{cases} \infty & \text{for nodes } i \text{ and } j \text{ not connected directly} \\ w_{ij} & \text{weight of connection link from nodes } i \text{ and } j \\ 0 & \text{for } i = j\text{, connecting a node with itself} \end{cases} \quad (3\text{-}22)$$

In some models, matrix elements for weighted networks are defined as

$$a_{ij} = \begin{cases} w_{ij} & \text{weight of connection link from nodes } i \text{ and } j \\ 0 & \text{when there is no direct link between nodes } i \text{ and } j \end{cases}, \quad (3\text{-}23)$$

The adjacency matrix is the key element for network topological analysis and generating it is the initial step in investigating the network characteristics.

3.7.2.2 Analyzing Network Topology. Key topological characteristics include the number of nodes (n), link number (E), average node degree ($\bar{K}$), characteristic path length (L), network diameter (D), and network clustering coefficient ($\bar{C}$). Node degree K_i is the number of direct connections each node has with their adjacent nodes. Average node degree is the mean node degree of all nodes and is expressed as follows:

$$\bar{K} = \frac{1}{n}\sum_{i=1}^{n} K_i \quad (3\text{-}24)$$

Network's diameter D is the maximum number of links to traverse along all possible paths in a network between nodes i and j. The local clustering coefficient of each node determines the tendency of its adjacent nodes to be connected. It is the fraction of the number of links between a node's adjacent nodes and their neighbors e_{ni} over the maximum possible number of links among the node:

$$C_i = \frac{2e_{ni}}{K_i(K_i - 1)} \tag{3-25}$$

The network clustering coefficient is the mean of local clustering coefficient and defined as follows:

$$\bar{C} = \frac{1}{n}\sum_{i=1}^{n} C_i \tag{3-26}$$

Global clustering coefficient C_G is the fraction of all closed triplets to the number of total open or closed triplets in a network.

The geodesic path is the shortest path between any two nodes in a network. The characteristic path length L is the mean over all geodesic path lengths d_{ij} for all possible pairs of nodes in a network and is denoted as follows:

$$L = \frac{1}{n(n-1)}\sum_{i \neq j} d_{ij} \tag{3-27}$$

3.7.2.3 Assessing Unweighted and Weighted Networks. The network efficiency or global network efficiency (E_G) of a metro network is a measure of how efficiently stations i and j connect within the network. The global efficiency of unweighted network is quantified as follows:

$$E_G = \frac{1}{n(n-1)}\sum_{i \neq j} \frac{1}{d_{ij}} \tag{3-28}$$

where n is the number of stations and d_{ij} is the geodesic path.

To create a weighted metro network, gathering data are required, and it depends on the network of interest. For instance, the dominant weight on transportation networks could be the ridership data or the flow of vehicles, the travel time, and so on. The weight w_{ij} assigned to each link could be achieved by normalizing the available data calculated as the ratio of weight of each link to the ridership of the link that carries the maximum flow as follows:

$$w_{ij} = \frac{W_{(i,j)}}{\max W_{(t,s)}} \tag{3-29}$$

where nodes i and j are any two random adjacent nodes connected by a link. Also, t and s are two neighboring stations belonging to the network located at two ends of a link that carries the maximum weight.

According to Saadat et al. (2019), the global efficiency of weighted network is similar to the unweighted network efficiency; however, the link weight on each geodesic path incorporates in the equation expressed as follows:

$$E_G = \frac{1}{n(n-1)} \sum_{i \neq j} \frac{W_{ij}}{d_{ij}} \tag{3-30}$$

where W_{ij} is the sum of all occurrences of w_{ij} on each geodesic path.

3.7.2.4 Assessing Vulnerability and Robustness. Robustness is the residual efficiency after a disruptive event. Robustness of an unweighted and weighted network can be quantitatively described by the changed connectivity after the removal of a network node. It is represented by Equations (3-12) and (3-13), respectively, as follows:

$$E_{Gi} = \frac{1}{n'(n'-1)} \sum_{i \neq j} \frac{1}{d'_{ij}} \tag{3-31}$$

$$E_{Gi} = \frac{1}{n'(n'-1)} \sum_{i \neq j} \frac{W'_{ij}}{d'_{ij}} \tag{3-32}$$

where n' is the number of nodes after the removal of failed nodes and d'_{ij} is the newly calculated geodesic path between nodes i and j. Also, W'_{ij} is the updated summation of weights on each geodesic path.

Global network efficiency is the basis of vulnerability assessment. Evaluating the network vulnerability is related to the changes in the global network efficiency because of disruptive events and can be quantified as follows:

$$V_i = \frac{E_G - E_{Gi}}{E_G} \tag{3-33a}$$

$$V = \max V_i \tag{3-33b}$$

where

V_i = Network vulnerability after disruption,

E_G = Initial global efficiency of a network prior to any disruption in the network, and

E_{Gi} = post global efficiency after removal of nodes or links in an infrastructure network.

3.7.2.5 Evaluating Resilience Metric for a Network. Examining changes to efficiency because of a disruptive event offers a basis for analyzing the resilience of a network. Improving network resilience helps enhance safety and requires appropriate resilience metrics. Resilience, defined as the persistence of performance under uncertainty because of a disruptive event, fundamentally, can be measured by defining performance, a performance loss profile because of disruption, and a recovery profile. The following two key measures are necessary to quantify network resilience as follows: (1) network performance before and after a disruptive event and (2) network performance recovery to the initial or some other level of functionality. The changes in the network performance constitute the resilience loss in the network (Bruneau and Reinhorn 2007, Henry and Ramirez-Marquez 2012).

Figure 3-17 is based on the resilience-triangle metric of Figure 3-1 that illustrates the concept of a resilience triangle, where Q_t on the vertical axis is the network performance. The shaded area presents the resilience loss expressed as follows:

$$R_e = \frac{\int_{t_0}^{t_0+t_h} Q_t dt}{t_h Q_0} \tag{3-34}$$

where
R_e = Resilience index,
Q_t = Performance of the disrupted system,

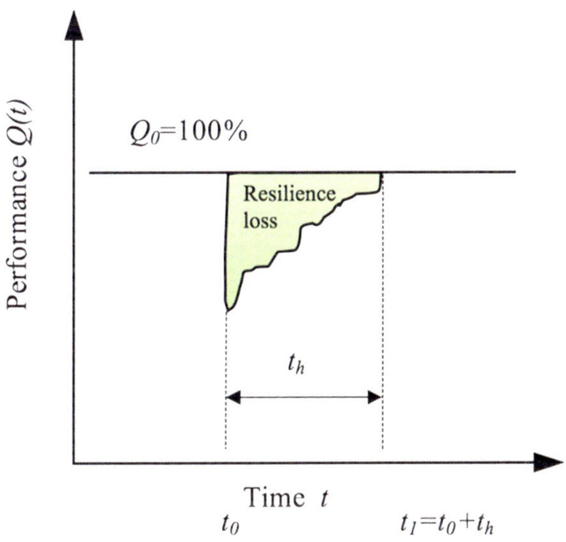

Figure 3-17. Resilience properties and resilience triangle in a system.

Q_0 = Performance before disruption,
t_0 = Time when the system is subjected to disruption, and
t_h = Period of time that the system needs to return to initial performance.

The global network efficiency could be used to represent the network performance. Hence, the resilience triangle for a metro network is

$$R_e = \frac{\int_{t_0}^{t_0+t_h} E_{Gt}\,dt}{t_h E_{G0}} \tag{3-35}$$

where E_{Gt} is the network efficiency at time t and E_{G0} is the initial efficiency of the metro network. During the time interval of t_h, the diminished performance system could be recovered to its initial performance level.

Table 3-8 summarizes the definitions and equations, previously presented, which are necessary to analyze the network topology, and Table 3-9 demonstrates the assessment metrics for a network that include efficiency, robustness, vulnerability, and resilience index.

3.7.3 Resilience of a Metro Network as an Example

The metro network of Washington, DC, is used as a case study. The Washington DC Metro network consists of six color-coded lines of red, green, yellow, blue, orange, and silver, as shown in Figure 3-18. Washington DC Metro is one of the busiest public transportation systems in the United States relative to the city's population.

To analyze the topology of the Washington DC Metro network, mapping the metro network into a topological graph is necessary. Figure 3-19 illustrates the topological graph of the Washington DC Metro network by assigning a number to each station.

Washington DC Metro consists of 91 stations (nodes) and 140 links. Saadat et al. (2019) analyzed the topological characteristic indicators of the metro, as demonstrated in Table 3-10.

For the purpose of analyses, the adjacency matrix $A_{ij} = \left[a_{ij}\right]_{91\times 91}$ for the Washington DC Metro network based on Equation (3-21) was developed. Results of subsequent topological analyses of the metro network are based on this adjacency matrix. Using Equation (3-24) and assuming that the network is unweighted, the average node degree $\bar{K}$ of the Washington DC Metro network is computed. The average node degree of the Washington DC Metro network equates to 2.0549. Thus, every station is linked to 2.0549 other stations. Based on the adjacency matrix, utilizing the algorithm developed by Floyd (1962) and using Equation (3-27), the characteristic

Table 3-8. Network Topological Characteristics.

Network Topological Characteristics	Definition	Equation
Number of nodes n	—	—
Number of links	—	—
Node degree K_i	Node degree K_i is the number of direct connections each node has with their adjacent nodes	—
Average node degree	Average node degree is the mean node degree of all nodes	$\bar{K} = \frac{1}{n}\sum_{i=1}^{n} K_i$
Geodesic path d_{ij}	Shortest path between nodes i and j	—
Characteristic path length	Average shortest path between any two nodes	$L = \frac{1}{n(n-1)}\sum_{i \neq j} d_{ij}$
Diameter of network D	Longest geodesic path in link count, among all network possible geodesic paths	—
Local clustering	Fractions of neighboring connections to node i to all possible links connected to node i	$C_i - \frac{2e_{ni}}{K_i(K_i - 1)}$
Network cluster coefficient	Average of local clustering of all nodes	$\bar{C} = \frac{1}{n}\sum_{i=1}^{n} C_i$

path length of the Washington DC Metro network is calculated. Equations (3-25) and (3-26) provide the clustering coefficient and efficiency of the network, respectively. The characteristic path length for the Washington DC Metro is 11.30 as the shortest path between any two stations, that is, a path needs to pass through 11.30 stations on the average. The longest path includes 28 stations defining the diameter of network. Regarding the Washington DC Metro clustering coefficient, because there are some nodes with node degrees of 1, its local clustering coefficient is not well defined. Thus, in this chapter, the global clustering coefficient is used to represent

Table 3-9. Network Assessment Metrics.

Network Assessment Metrics	Definition	Equation
Unweighted network efficiency	An indicator of node connectivity that is proportional to the reciprocal of shortest distance	$E_G = \frac{1}{n(n-1)} \sum_{i \neq j} \frac{1}{d_{ij}}$
Weighted network efficiency	An indicator of node connectivity that is proportional to the summation of weights on each shortest path reciprocal of the shortest distance	$E_G = \frac{1}{n(n-1)} \sum_{i \neq j} \frac{W_{ij}}{d_{ij}}$
Robustness for unweighted network	Robustness is the residual efficiency after a disruptive event	$E_{Gi} = \frac{1}{n'(n'-1)} \sum_{i \neq j} \frac{1}{d'_{ij}}$
Robustness for weighted network	Robustness is the residual efficiency after a disruptive event	$E_{Gi} = \frac{1}{n'(n'-1)} \sum_{i \neq j} \frac{W'_{ij}}{d'_{ij}}$
Vulnerability	Vulnerability relates to the changes on the global network efficiency. Vulnerability of each node is indicated by V_i and V *is* the network vulnerability	$V_i = \frac{E_G - E_{Gi}}{E_G}$ $V = \max V_i$
Resilience index	Resilience index shows how the diminished performance system could be recovered to its initial performance level during the time interval of t_h	$R_e = \frac{\int_{t_0}^{t_0+t_h} E_{Gt}\,dt}{t_h E_{G0}}$

the Washington DC Metro network average clustering coefficient. The global clustering coefficient is 0.038.

To assess the unweighted Washington DC Metro efficiency, all geodesic paths between any two nodes using its adjacency matrix and the Floyd (1962) algorithm are evaluated to produce an efficiency of 0.14320 for a network without any node or link failures. The initial metro network efficiency is the basis of calculating network vulnerability and resilience because of node or link failures. Here it should be noted that the Floyd

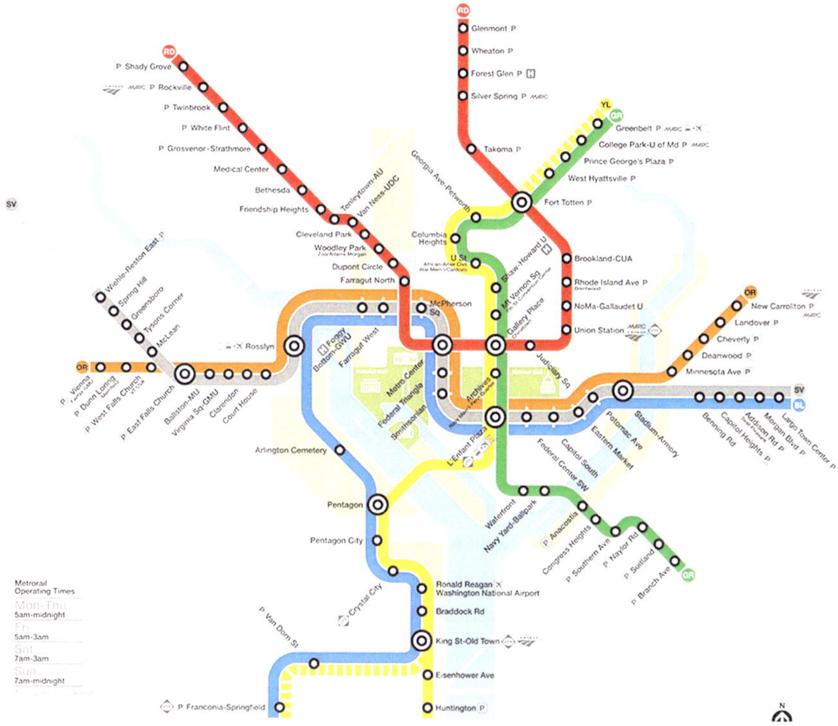

Figure 3-18. Washington DC Metro map.
Source: Washington DC Metro Fiscal Year Budget (2018).

(1962) algorithm is only applicable for L-space networks. By definition, in an L-space graph, there is only one link to connect any two consecutive nodes. The Washington DC Metro network is an L-space network.

The vulnerability assessment of the Washington DC Metro network covers two cases: (1) node removal one at a time and (2) link removal one at a time. For both cases, Equations (3-33a) and (3-33b) are used to calculate the vulnerability. In Equation (3-33a), E_{Gi} refers to the efficiency of a network after a node removal or a link removal, providing measures of respective robustness. Characterizing the node connectivity is the key element for measuring robustness, which is a necessary pursuit for vulnerability assessment. To calculate E_{Gi}, Equation (3-31) is used with a new set of d_{ij} after node or link removal, which also represents the robustness of the network. The size of the adjacency matrix after one node removal is 90 by 90. Figure 3-20 displays the vulnerability profile of the metro network when subjected to node removal for all 91 nodes. Also, Table 3-11 lists the 10 most critical stations along with their vulnerability values.

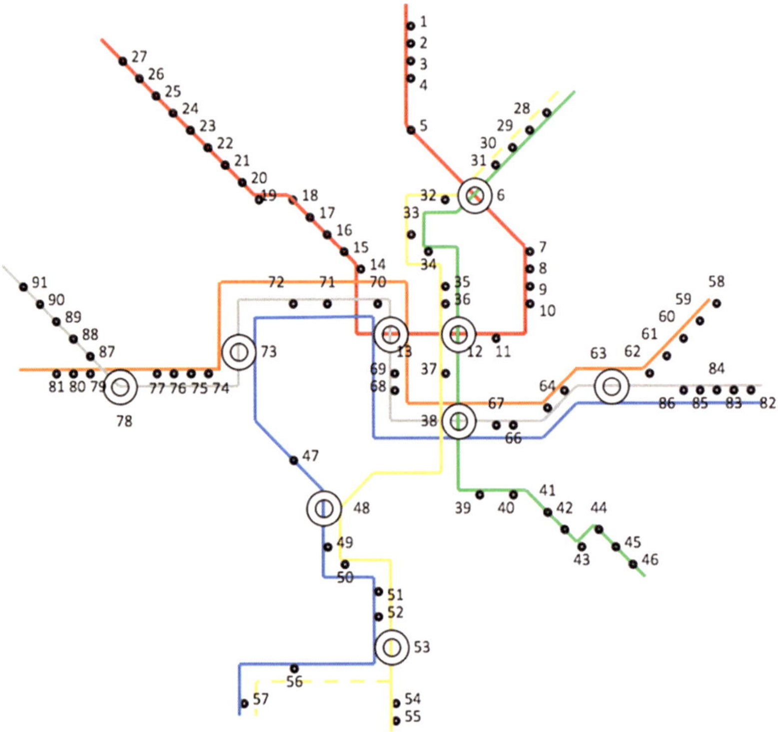

Figure 3-19. Topological graph of Washington DC Metro network.

Equation (3-33b) equates the network vulnerability to the maximum vulnerability among all nodes removed once at a time. Thus, the vulnerability of the network is 0.30 and L' Enfant Plaza Station is of the greatest vulnerability in the Washington DC Metro network. By removing the L' Enfant Plaza Station, the efficiency of the network would be drastically reduced by 29.78%.

Using the link removal method, the regenerated network is developed, after which the associated network efficiency is derived using Equation (3-31). Measuring network efficiency in this case is based on the Floyd (1962) algorithm by changing the adjacency matrix after each link removal and not changing the size of the adjacency matrix. Table 3-12 shows the vulnerability measurements for 15 most critical links in the Washington DC Metro network. In Table 3-12, (i,j) is an indicator for a link between node i and node j.

Based on Table 3-12, the most critical lines in the Washington DC Metro network are the blue–orange–silver lines, which share the same track and, therefore, are considered as one single-use link. The most vulnerable

Table 3-10. Initial Characteristics of Washington, DC Metro Network.

Characteristics of Metro Network	Calculated Value for the Washington, DC Metro	Interpretations
Number of nodes n	91	91 Stations
Number of links	140	93 Links with lines sharing the same tracks treated as a singly link
Average node degree $\bar{K} = \frac{1}{n}\sum_{i=1}^{n} K_i$	2.0549	Each station averagely connects to 2.0549 other station
Characteristic path length $L = \frac{1}{n(n-1)}\sum_{i \neq j} d_{ij}$	11.30	The average shortest path between any two stations is 11.30 links
Diameter of network D	28	The longest geodesic path in link count, among all network possible geodesic paths is 28
Network cluster coefficient $\bar{C} = \frac{1}{n}\sum_{i=1}^{n} C_i$	0.038*	The average of local clustering of all nodes as the fractions of neighboring connections to node s_i to all possible links connected to node s_i

*This measurement is based on using global clustering coefficient because of the absence of well-defined local clustering coefficient in Washington, DC Metro network.

segment of these lines is where they connect east Washington DC to the downtown area. The portions of orange–silver lines connecting western Washington DC to downtown and the blue–yellow lines connecting the southern part of the city to the downtown area are vulnerable as well. In addition, the red line where several links connect the northwest quadrant

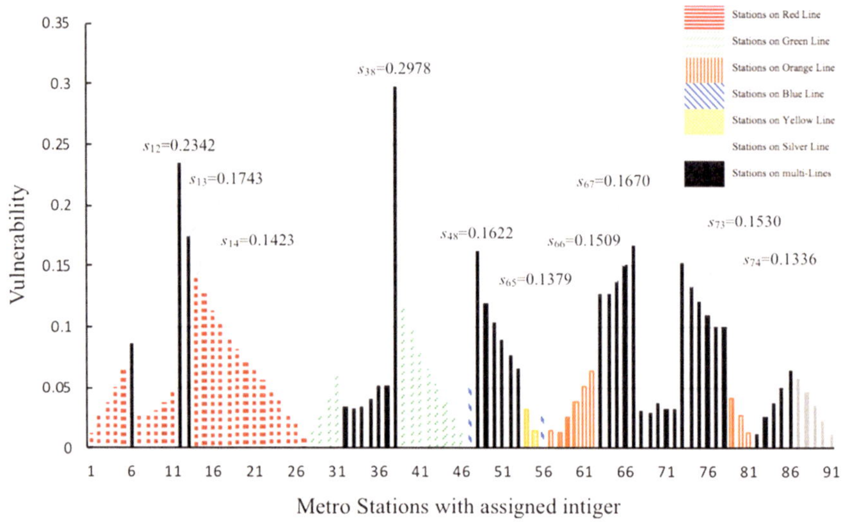

Figure 3-20. Vulnerability of Washington DC Metro network subjected to node removal for all 91 stations.

Table 3-11. Ten Most Critical Nodes to the Network Vulnerability of Washington, DC Metro.

Rank Order	Name of the Station	Station Number	Node Degree	Vulnerability Magnitude (%)
1	L' Enfant Plaza	38	5	29.78
2	Gallery Place	12	4	23.42
3	Metro Center	13	4	17.43
4	Federal Center SW	67	2	16.70
5	Pentagon	48	3	16.22
6	Rosslyn	73	3	15.30
7	Capital South	66	2	15.09
8	Farragut North	14	2	14.23
9	Eastern Market	65	2	13.79
10	Court House	74	2	13.36

Table 3-12. Fifteen Most Critical Links for the Network Vulnerability of Washington, DC Metro.

Rank Order	Link (*i,j*)	Station Names at Two Ends of Each Link	Color-Coded Lines	Vulnerability Magnitude (%)
1	(13,69)	(Metro Center, Federal Triangle)	Blue–orange–silver	16.07
2	(68,69)	(Smithsonian, Federal Triangle)	Blue–orange–silver	14.42
3	(38,68)	(L′ Enfant Plaza, Smithsonian)	Blue–orange–silver	13.14
4	(73,74)	(Rosslyn, Court House)	Orange–silver	12.78
5	(38,67)	(L′ Enfant Plaza, Federal Center SW)	Blue–orange–silver	12.07
6	(13,14)	(Metro Center, Farragut North)	Red	11.71
7	(48, 49)	(Pentagon, Pentagon City)	Blue–yellow	11.57
8	(74,75)	(Court House, Clarendon)	Orange–silver	11.50
9	(63,64)	(Stadium Armory, Potomac Ave)	Blue–orange–silver	11.14
10	(75,76)	(Clarendon, Virginia-Square-GMU)	Orange–silver	10.42
11	(14,15)	(Farragut North, Dupont Circle)	Red	10.14
12	(49,50)	(Pentagon City, Crystal City)	Blue–yellow	9.85
13	(76,77)	(Virginia-Square-GMU, Ballston-MU)	Orange–silver	9.50
14	(15,16)	(Dupont Circle, Woodley Park)	Red	8.86
15	(38,39)	(L′ Enfant Plaza, Waterfront)	Green	8.78

of Washington, DC, to the downtown area and a segment of the green line responsible for connecting the southeast quadrant of the city to the downtown area are considered the most vulnerable links. Notably, in analyzing link removal, the segments connected to one of the transfer stations are exposed to potential threats and are ranked among the most critical links in the Washington DC Metro network.

Vulnerability assessment is the basis for quantifying the resilience of the network. To evaluate the resilience index, incorporating the time (t_h) into Equation (3-35) is required for the efficient recovery of the disrupted network to its initial state.

3.7.4 Recovery Strategies of Networked Infrastructure

Regional resilience analysis is about quantifying the ability of structures, infrastructure, and communities to absorb disturbances from external stressors and return to normalcy (Bruneau et al. 2003; Sharma et al. 2018, 2019a; Gardoni and Murphy 2018; Gardoni 2019). Resilient infrastructure is cardinal to prompt societal recovery and enhanced regional resilience (Tabandeh et al. 2019).

To promote resilient communities, decisions on risk mitigation and recovery management should be informed by regional resilience analysis. The resilience analysis of infrastructure broadly consists of (1) modeling the postdisruption physical recovery, (2) modeling the associated service recovery as the physical recovery progresses, and (3) predicting the implication on desired resilience objectives (Sharma et al. 2019a, b). Furthermore, integrating the resilience analysis with an optimization formulation can aid to unveil the bottlenecks of regional resilience and guide the development of effective mitigation and recovery strategies (Sharma et al. 2019b). Although the focus herein is on physics-based recovery formulations, empirical models have been developed for specific infrastructure subject to specific disruptive events (Reed et al. 2009, 2015). Such models are typically calibrated using collected data from past similar events or physics-based formulations.

3.7.4.1 Physical Recovery Modeling. Physical recovery modeling aims to develop a detailed schedule for the repair or replacement of damaged infrastructure components (e.g., electric power transformers) and model the effects of the physical progression on the structural characteristics. A significant challenge is to schedule a large number of recovery activities for the repair of damaged components that are spread over a large area. The development of a realistic recovery schedule requires capturing the physical and logical constraints such as the availability of recovery crews, logical precedence of recovery activities, and work continuity (Sharma et al. 2019a).

To develop a realistic recovery schedule and overcome computational complexities, we use a multiscale approach for recovery modeling (Sharma et al. 2019a) with two levels of hierarchy, named as zonal and local recovery scales. At the zonal scale, a set of recovery zones is defined that partitions infrastructure based on functional association and geographic proximity of components, land use, or community neighborhoods. The components that are damaged in each zone are assumed to recover with the same zonal priority, which can be decided at the higher levels of management. At the local scale, the required recovery activities in each zone are identified and assigned to available crews. The details of local schedules and securing the required resources can be decided in coordination with local authorities.

For a developed recovery schedule, we can model the effects on the structural characteristics of each component in terms of its state variables that define the components (Sharma et al. 2018, 2019a). Examples of such state variables include material properties, member dimensions, and boundary conditions. We then use the estimates of state variables for service recovery modeling, discussed next.

3.7.4.2 Service Recovery Modeling. To accurately model the service recovery, it is essential to capture the service flow stability in addition to the physical connectivity of infrastructure. This is because service areas that are physically connected to infrastructure may still remain nonfunctional if specific flow constraints are violated (e.g., voltage stability in electric power infrastructure). The mathematical representation of infrastructure consists of a collection of networks, where each network captures a specific performance of the infrastructure (Sharma and Gardoni 2019). For example, we model the electric power infrastructure with two networks, called structural and power flow networks. The structural network captures the structural performance, and the power flow network captures the respective functionality performance. Recognizing the complex, multidisciplinary nature of the service recovery modeling, we decouple (inter)dependent networks such that each network can be modeled in a rigorous and consistent manner (Sharma and Gardoni 2019). We then separately capture the effects of (inter)dependencies through a set of interface functions that modify the base estimates of the performance measures of each network.

Let $G = \{G^{[k]}:k = 1, \ldots, K\}$ be the collection of all networks required to model the service recovery of interdependent infrastructure. Each network $G^{[k]}$ consists of a set of vertices and a set of edges (Guidotti et al. 2017, Sharma and Gardoni 2019). The vertices are nodal components (e.g., power generators) and the edges are line components (e.g., power transmission lines). Each network is characterized by a unique set of capacity, demand, and supply measures (Sharma and Gardoni 2019). For example, one can define capacity measures for a power transformer relative to its structural failure modes and relative to its functionality as the maximum power flow. For each capacity measure, there is a respective demand measure owing to the occurrence of an extreme event or regular service conditions. Furthermore, the supply measure captures the ability of the network to serve the demand placed on its components (Sharma and Gardoni 2019).

Service recovery modeling builds upon the recovery models for the capacity, demand, and supply measures of all networks in G. To model the recovery of capacity and demand measures for each network, we use the estimates of the state variables for each component in the respective capacity and demand models (Gardoni et al. 2002, 2003). Following Sharma and Gardoni (2019), we can develop a set of equations $\mathbf{S}^{[k]}(\tau)$ to model the supply measure of network $G^{[k]}$ at any time τ since the beginning of the recovery. In general, $\mathbf{S}^{[k]}(\tau)$ is a function of a vector of control state variables (e.g., voltage in power flow network), a vector of capacity estimates for the components of $G^{[k]}$, a vector of demand estimates, and a vector of model parameters. For structural networks, the supply estimate for each component is simply the same as the imposed demand insofar because the demand has not exceeded the respective capacity; otherwise, in the case of a structural failure, the supply estimate becomes zero. For flow networks, the model for $\mathbf{S}^{[k]}(\tau)$ represents a flow analysis (Sharma et al. 2019a). We can then define derived performance measures $\mathbf{Q}^{[k]}(\tau)$ as functions of the capacity, demand, and supply estimates to capture other aspects of service recovery. Examples of $\mathbf{Q}^{[k]}(\tau)$ include the reliability of components in a structural network (Sharma et al. 2018) and the fraction of demand that can be supplied in a flow network (Sharma et al. 2019a).

To account for the effects of (inter)dependencies among different networks, Sharma and Gardoni (2019) introduced a set of interface functions that modify the base estimates of capacity and demand measures. For example, the interface functions modify the capacity and demand estimates of components in a power flow network as functions of the reliability estimates of components in the associated structural network. We can then obtain the modified supply estimates using the modified capacity and demand estimates in the model for the supply measure. Likewise, we can obtain the modified estimates of the derived performance measures $\mathbf{Q}'^{[k]}(\tau)$. For the regional resilience analysis, we define the recovery curve in terms of an aggregate performance measure $S'^{[\mathrm{agg}]}(\tau)$ that combines the availability of all services provided by interdependent infrastructure (Sharma et al. 2019a).

3.7.4.3 Resilience Quantification for Recovery Purposes. Once properly developed, recovery curves provide complete information about resilience for the purpose of quantification (Bruneau et al. 2003, Chang and Shinozuka 2004, Bruneau and Reinhorn 2007, Reed et al. 2009,

Cimellaro et al. 2010, Decò et al. 2013, Ayyub 2014a, Sharma et al. 2018). Section 3.3 provides fundamental models for quantifying resilience. Recently, Sharma et al. (2018) provides a computational approach for resilience quantification that decomposes the recovery curve in terms of all its partial descriptors. These resilience metrics are defined in analogy with the definitions of the statistical moments in probability theory. Examples of these resilience metrics that are equivalent to central tendency and dispersion measures are as follows: center of resilience, ρ_Q, that combines the residual performance in the immediate aftermath of a disruption with the recovery duration. We can write ρ_Q in analogy with the mean of a random variable as

$$\rho_Q = \frac{\int_0^{T_R} \tau dQ(\tau)}{\int_0^{T_R} dQ(\tau)} \tag{3-36}$$

In addition, resilience bandwidth, χ_Q, is a measure of dispersion of the recovery with χ_Q provided in analogy with the standard deviation of a random variable as

$$\chi_Q = \sqrt{\frac{\int_0^{T_R} (\tau - \rho_Q)^2 dQ(\tau)}{\int_0^{T_R} dQ(\tau)}} \tag{3-37}$$

The definition of these resilience metrics is general, and one can systematically extend any subset of metrics with additional ones to quantify resilience with desired accuracy. Furthermore, the resilience metrics are simple and have tangible interpretations that facilitate the communication of resilience between researchers and decision makers.

3.7.4.4 Regional Recovery Optimization. Regional recovery optimization aims to set the priority of the recovery zones for each damaged network (i.e., decision variables) such that the disrupted services are restored as fast as possible, while minimizing the incurred cost (Sharma et al. 2019b). The multiscale approach to recovery modeling (Sharma et al. 2019a) aids to overcome the computational complexity of recovery optimization. The high-dimensional optimization problem of scheduling individual recovery activities has been broken down into a hierarchy of decoupled low-dimensional optimization problems at the zonal and local scales.

Given a collection of damaged networks (made up of a set of damaged nodal components and a set of damaged line components), we can subdivide each network into different recovery zones. We can then write the zonal-scale optimization problem as the minimization of a set of objective function(s) (e.g., resilience metrics or cost) obtained by considering different priorities of the recovery zones.

The zonal-scale optimization problem is subject to physical and logical scheduling constraints for the recovery activities within each zone and service recovery constraints. Each set of these constrains entails a nested optimization. To schedule the recovery activities within each zone, we formulate a local-scale optimization problem whose objective is to minimize the recovery duration within the zone, while complying with physical and logical constraints to implement the recovery schedule. The objective of the service recovery optimization is to minimize a measure of discrepancy between the loss function estimates of the demand and supply measures through the recovery, while complying with the network-specific constraints, such as power balance equations for the power flow network (Glover et al. 2012). Further details about the recovery optimization can be found in Sharma et al. (2019b).

3.7.4.5 Resilience-Informed Infrastructure Recovery Example. The definition of an optimal recovery strategy is illustrated by modeling the performance of the electric power infrastructure in Shelby County, Tennessee. Shelby County has an approximately 1,000,000 population, and the region is subject to seismic hazards originating from New Madrid Seismic Zone (NMSZ). In this example, we consider a historical scenario earthquake with a moment magnitude of 7.7 and the epicenter at 35.93°N and 89.92°W. We model the spatial variation of the earthquake intensity measures by using a three-dimensional physics-based model to capture near-field effects [Guidotti, R., S. Tian, and P. Gardoni, "Simulation of seismic wave propagation in the Metro Memphis Statistical Area (MMSA)," in preparation] and ground motion prediction equations for far-field regions (Steelman et al. 2007).

The electrical power infrastructure in Shelby County is managed by the Memphis Light, Gas, and Water (MLGW) Division. The balancing authority of the region is the Tennessee Valley Authority (TVA) who also owns and operates the generators and transmission lines providing power to MLGW. The model for the power flow analysis is provided by Sharma et al. (2019a), building on the information provided in Shinozuka et al. (1998) and Birchfield et al. (2017). Figure 3-21 shows the topology of the developed model for Shelby County (b) and Tennessee (a).

To model the physical recovery, we estimate the damage to the vulnerable components and develop a detailed recovery schedule for the

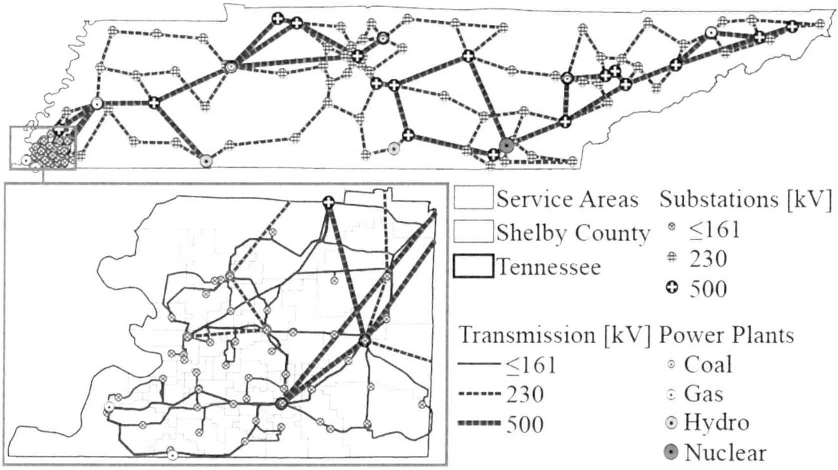

Figure 3-21. Electric power infrastructure in (a) Tennessee, and (b) Shelby County.
Source: Adapted from Sharma et al. (2019a).

repair or replacement of damaged components. Transformers, circuit breakers, and disconnect switches are vulnerable components to seismic excitations, which are all located in electric power substations. To develop the recovery schedule, we consider each substation and the corresponding service area as a single recovery zone. Given that two different agencies manage the electric power infrastructure inside and outside Shelby County, we define four different recovery projects as follows: (1) MLGW critical repairs, for nonfunctional substations in Shelby County; (2) MLGW noncritical repairs, for functional but damaged substations in Shelby County; (3) TVA critical repairs, for nonfunctional substations in Tennessee; and (4) TVA noncritical repairs, for functional but damaged substations in Tennessee. Further details of the recovery schedule can be found in Sharma et al. (2019a).

To model service recovery, we develop a structural network, $G^{[1]}$, and a power flow network, $G^{[2]}$. The structural capacity and demand measures are in terms of the hazard intensity measure, whereas the flow capacity and demand measures are in terms of the apparent power. The capacity of the flow network is dependent on the structural network. We account for this dependency in the modified flow capacity estimates and accordingly obtain the modified flow supply estimates by running power flow analyses. To summarize the overall service recovery, we define the aggregate performance measure $Q'^{[\text{agg}]}(\tau) = \sum_{\text{cell}=1}^{n_{\text{cell}}} w_{\text{cell}} Q'^{[2]}_{\text{cell}}(\tau)$, where cell is the service area of each substation; w_{cell} is a weight for the recovery cell that is proportional to the power demand in the recovery cell; and $Q'^{[2]}_{\text{cell}}(\tau)$ is the fraction of the power demand that is supplied in the recovery *cell*.

The scenario earthquake is estimated to cause damage to components in 17 of the 36 zones managed by MLGW, which require critical repairs. In this example, we use ρ_Q as the sole objective function in recovery optimization. To solve the optimization problem, we use a genetic algorithm (Goldberg 1989), whereas other algorithms could be used as well. Figure 3-22 shows the results for the service recovery in terms of $Q'^{[2]}_{\text{cell}}(\tau)$, which is a binary-value quantity (dark gray is nonfunctional, and light gray is functional). Figure 3-22(a) shows the results according to the current recovery practice as defined in MLGW (2017), and Figure 3-22(b) shows the results according to the optimized recovery schedule. In Figure 3-22(a), we observe that $Q'^{[2]}_{\text{cell}}(\tau)$ for some recovery cells fluctuates over time. This is because as the recovery advances, redistribution of loads on operating buses can result in voltage collapse. The optimized recovery results in $\rho_Q = 18.1$ h, compared with $\rho_Q = 26.5$ h for the current recovery practice (i.e., a 30.2% improvement). We can also observe that the improvement in $Q'^{[2]}_{\text{cell}}(\tau)$ is not uniform across the region because some areas experience slower recovery than the others. This is because the focus of ρ_Q, as the recovery objective, is on the recovery duration, thus not capturing the temporal and spatial variabilities in the recovery. Instead, one can use the formulation in Section 3.9.4 to define a multiobjective optimization problem that captures all desired resilience objectives in developing the recovery schedule.

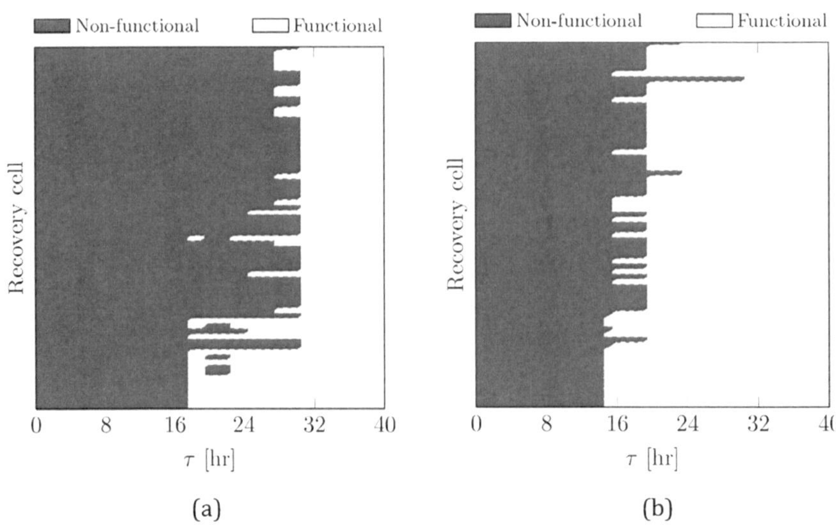

Figure 3-22. Predicted performance of the electric power infrastructure under (a) current recovery practice, and (b) optimized recovery schedule.

REFERENCES

AAR (Association of American Railroads). 2010. *Class I railroad statistics*. Washington, DC: AAR. Accessed July 10, 2019. http://www.aar.org.

AAR. 2018. *140,000-Mile private rail network delivers for America's economy*. Washington, DC: AAR. Accessed July 10, 2019. http://archive.freightrailworks.org/network/class-i/.

Agusdinata, B. 2008. *Exploratory modeling and analysis*. Delft, Netherlands: TU Delft (Delft University of Technology).

Alderson, D., G. Brown, and W. Carlyle. 2015. "Operational models of infrastructure resilience, risk analysis." *Risk Anal.* 35 (4): 562–586.

Almufti, I., and M. Willford. 2014. "The REDi™ rating system: A framework to implement resilience-based earthquake design for new buildings." In *Proc., 10th U.S. National Conf. on Earthquake Engineering*, Earthquake Engineering Research Institute, Anchorage, Alaska. Accessed March 14, 2020. https://datacenterhub.org/resources/12400/download/10NCEE-001055.pdf.

Ayyub, B. M. 2014a. "Systems resilience for multi-hazard environments: Definition, metrics and valuation for decision making." *Risk Anal.* 34 (2): 340–355.

Ayyub, B. M. 2014b. *Risk analysis in engineering and economics*. 2nd ed. Boca Raton, FL: Chapman & Hall/CRC.

Ayyub, B. 2015. "Practical resilience metrics for planning, design and decision making." *J. Risk Uncertainty Eng. Syst. Part A: Civ. Eng.* 1 (3): 04015008.

Bankes, S. 1993. "Exploratory modeling for policy analysis." *Oper. Res.* 41 (3): 435–449.

Batouli, M., and A. Mostafavi. 2014. "A hybrid simulation framework for integrated management of infrastructure networks." In *Proc., Winter Simulation Conf.*

Batouli, M., and A. Mostafavi. 2016. "Assessment of sea-level rise adaptations in coastal infrastructure systems: Robust decision making under uncertainty." In *Proc., Construction Research Congress*. 2016, San Juan, Puerto Rico, May 31–June 2, 2016, 1455-1464. https://ascelibrary.org/doi/book/10.1061/9780784479827

Batouli, M., and A. Mostafavi. 2018. "Multiagent simulation for complex adaptive modeling of road infrastructure resilience to sea-level rise." *Comput.-Aided Civ. Infrastruct. Eng.* 33 (5): 393–410.

Bhamidipati, S., T. van der Lei, and T. Herder. 2016. "A layered approach to model interconnected infrastructure and its significance for asset management." *Eur. J. Transp. Infrastruct. Res.* 16 (1): 254–272.

Birchfield, A. B., K. M. Gegner, T. Xu, K. S. Shetye, and T. J. Overbye. 2017. "Statistical considerations in the creation of realistic synthetic power grids for geomagnetic disturbance studies." *IEEE Trans. Power Syst.* 32 (2): 1502–1510.

Biringer, B., E. D. Vugrin, and D. Warren. 2013. *Critical infrastructure system security and resilience*. Boca Raton, FL: CRC Press.

Boardman, J., and B. Sauser. 2006. "System of systems—The meaning of OF." In *Proc., IEEE/SMC Int. Conf. on System of Systems Engineering.*, Los Angeles, 118–123. doi:10.1109/SYSOSE.2006.1652284 New York: IEEE.

Bruneau, M., S. E. Chang, R. T. Eguchi, G. C. Lee, T. D. O'Rourke, A. M. Reinhorn, et al. 2003. "A framework to quantitatively assess and enhance the seismic resilience of communities." *Earthquake Spectra* 19 (4): 733–752.

Bruneau, M., and A. Reinhorn. 2007. "Exploring the concept of seismic resilience for acute care facilities." *Earthquake Spectra* 23 (1): 41–62.

Chang, S. E., and M. Shinozuka. 2004. "Measuring improvements in the disaster resilience of communities." *Earthquake Spectra* 20 (3): 739–755.

Cimellaro, G. P., A. M. Reinhorn, and M. Bruneau. 2010. "Framework for analytical quantification of disaster resilience." *Eng. Struct.* 32 (11): 3639–3649.

Decò, A., P. Bocchini, and D. M. Frangopol. 2013. "A probabilistic approach for the prediction of seismic resilience of bridges." *Earthquake Eng. Struct. Dyn.* 42 (10): 1469–1487.

Dehghani, M. S., G. Flintsch, and S. McNeil. 2014. "Impact of road conditions and disruption uncertainties on network vulnerability." *J. Infrastruct. Syst.* 20 (3): 04014015.

DeLaurentis, D. 2005. "Understanding transportation as a system-of-systems design problem." In *Proc., 43rd AIAA Aerospace Sciences Meeting and Exhibit*. Reston, VA: American Institute of Aeronautics and Astronautics.

DeLaurentis, D., and R. K. C. Callaway. 2004. "A system-of-systems perspective for public policy decisions." *Rev. Policy Res.* 21 (6): 829–837.

Field, C., R. Look, and T. Lindsay. 2016. "Resilience insight: 12 cities assessment. BuroHappold Engineering, BRE and Worshipful Company of Constructors." Accessed May 21, 2020. https://www.burohappold.com/wp-content/uploads/2016/06/2016-Royal-Charter-International-Research-Award-BuroHappold-Resilience-Insight-12-Cities-Assessment-v2.pdf.

Field, C., R. Look, and T. Lindsay. 2017. "A comprehensive approach to city & building resilience." In *Proc., AEI Annual Convention*. Accessed March 14, 2020. https://www.researchgate.net/publication/315850846_A_Comprehensive_Approach_to_City_and_Building_Resilience.

Fisher, R., and M. Norman. 2010. "Developing measurement indices to enhance protection and resilience of critical infrastructures and key resources." *J. Bus. Continuity*, 4(3), 191–206.

Floyd, R. W. 1962. "Algorithm 97: Shortest path." *Commun. ACM* 5 (6): 345–345.

Gao, J., B. Barzel, and A.-L. Barabási. 2016. "Universal resilience patterns in complex networks." *Nature* 530 (7590): 307.

Gardoni, P., ed. 2019. *Handbook of sustainable and resilient infrastructure*. Abington, UK: Routledge.

Gardoni, P., A. Der Kiureghian, and K. M. Mosalam. 2002. "Probabilistic capacity models and fragility estimates for reinforced concrete columns based on experimental observations." *J. Eng. Mech.* 128 (10): 1024–1038.

Gardoni, P., K. M. Mosalam, and A. Der Kiureghian. 2003. "Probabilistic seismic demand models and fragility estimates for RC bridges." *J. Earthquake Eng.* 7 (spec01): 79–106.

Gardoni, P., and C. Murphy. 2018. "Society-based design: promoting societal well-being by designing sustainable and resilient infrastructure." *Sustainable Resilient Infrastruct.* 5(1-2), https://doi.org/10.1080/23789689.2018.1448667

Gilbert, S., and B. M. Ayyub. 2016. "Models for the economics of resilience." *J. Risk Uncertainty Eng. Syst. Part A: Civ. Eng.* 2 (4): 04016003.

Glover, J. D., M. S. Sarma, and T. Overbye. 2012. *Power system analysis and design*. Boston: Cengage Learning.

Goldberg, D. E. 1989. *Genetic algorithms in search, optimization, and machine learning*. Reading, MA: Addison-Wesley.

Guidotti, R., P. Gardoni, and Y. Chen. 2017. "Network reliability analysis with link and nodal weights and auxiliary nodes." *Struct. Saf.* 65: 12–26.

Hassan, E., and H. Mahmoud. 2019. "Full functionality and recovery assessment framework for a hospital subjected to a scenario earthquake event." *Eng. Struct.* 188: 165–177.

Henry, D., and J. E. Ramirez-Marquez. 2012. "Generic metrics and quantitative approaches for system resilience as a function of time." *Reliab. Eng. Syst. Saf.* 99: 114–122.

Jenelius, E., and L.-G. Mattsson. 2015. "Road network vulnerability analysis: Conceptualization, implementation and application." *Comput. Environ. Urban Syst.* 49 (1): 136–147.

Jenelius, E., T. Petersen, and L.-G. Mattsson. 2006. "Importance and exposure in road network vulnerability analysis." *Transp. Res. Part A Policy Pract.* 40 (7): 537–560.

Jones, D. A., L. K. Nozick, M. A. Turnquist, M. W. Hollingsworth, C. R. Lawton, M. A. Ehlen, et al. 2003. *Impact analysis of potential disruptions to major railroad bridges in the U.S.—Phase I Report*. Albuquerque, NM: National Infrastructure Simulation and Analysis Center, Sandia National Laboratories.

Kaplan, S., and B. J. Garrick. 1981. "On the quantitative definition of risk." *Risk Anal.* 1: 11–27.

Koetse, M. J., and P. Rietveld. 2009. "The impact of climate change and weather on transport: An overview of empirical findings." *Transp. Res. Part D Transp. Environ.* 14 (3): 205–221.

Kwakkel, J. H., and E. Pruyt. 2013. "Exploratory modeling and analysis, an approach for model-based foresight under deep uncertainty." *Technol. Forecasting Social Change* 80 (3): 419–431.

Lambert, J. H., Y.-J. Wu, H. You, A. Clarens, and B. Smith. 2013. "Climate change influence on priority setting for transportation infrastructure assets." *J. Infrastruct. Syst.* 19 (1): 36–46.

Latora, V., and M. Marchiori. 2001. "Efficient behavior of small-world networks." *Phys. Rev. Lett.* 87 (19): 4.

Lempert, R., N. Nakicenovic, D. Sarewitz, and M. Schlesinger. 2004. "Characterizing climate-change uncertainties for decision-makers. An editorial essay." *Clim. Change* 65 (1–2): 1–9.

Maier, M. W. 1998. "Architecting principles for systems-of-systems." *Syst. Eng.* 1 (4): 267–284.

MLGW (Memphis Light, Gas and Water). 2017. "How MLGW restores power." Press release. Accessed July 10, 2019. http://www.mlgw.com/news/how-mlgw-restores-power-5-2017.

Mori, Y., and B. R. Ellingwood. 1993. "Time-dependent system reliability analysis by adaptive importance sampling." *Struct. Saf.* 12 (1): 59–73.

Mostafavi, A., D. Abraham, and D. DeLaurentis. 2014. "Ex-ante policy analysis in civil infrastructure systems." *J. Comput. Civil Eng.* 28 (5): A4014006.

Mostafavi, A., D. M. Abraham, D. DeLaurentis, and J. Sinfield. 2011. "Exploring the dimensions of systems of innovation analysis: A system of systems framework." *IEEE Syst. J.* 5 (2): 256–265.

Mostafavi, A., D. Abraham, D. DeLaurentis, J. Sinfield, A. Kandil, and C. Queiroz. 2016. "Agent-based simulation model for assessment of financing scenarios in highway transportation infrastructure systems." *J. Comput. Civil Eng.* 30 (2): 04015012.

Mostafavi, A., D. M. Abraham, and J. Lee. 2012. "System-of-systems approach for assessment of financial innovations in infrastructure." *Built Environ. Project Asset Manage.* 2 (2): 250–265.

Mostafavi, A., and N. E. Ganapati. 2019. "Toward convergence disaster research: Building integrative theories using simulation." *Risk Anal.* https://doi.org/10.1111/risa.13303

NEI (Nuclear Energy Institute). 2016. *Diverse and flexible coping strategies (FLEX) implementation guide.* NEI 12-06, Revision 4. Washington, DC: NEI. Accessed March 18, 2020. https://www.nrc.gov/docs/ML1635/ML16354B421.pdf.

NEII (Nuclear Energy Institute Infographic). 2012. "Making safe nuclear energy safer after Fukushima." Accessed March 18, 2020. http://assets.safetyfirst.nei.org.s3.amazonaws.com/wordpress/wp-content/uploads/2012/03/NEI_Infographic.jpg.

Newman, M. E. J. 2010. *Mathematics of networks. Networks, an introduction*, 99–166. New York: Oxford University Press.

Ortiz-García, J. J., S. B. Costello, and M. S. Snaith. 2006. "Derivation of transition probability matrices for pavement deterioration modeling." *J. Transp. Eng.* 132 (2): 141–161.

Pahl-Wostl, C. 2002. "Towards sustainability in the water sector—The importance of human actors and processes of social learning." *Aquat. Sci.* 64 (4): 394–411.

Paul, S. K. 2014. "Vulnerability concepts and its application in various fields: A review on geographical perspective." *J. Life Earth Sci.* 8: 63–81.

Petit, T. J., J. Fiksel, and K. L. Croxton. 2010. "Ensuring supply chain resilience: Development of a conceptual framework." *J. Bus. Logist.* 31: 1–21.

Powell, M., J. Taylor, and S. Baier. 2013. "Developing the FLEX Plan, 2013." NS Energy. Accessed March 14, 2020. https://www.nsenergybusiness.com/features/featuredeveloping-the-flex-plan/.

PPD (Presidential Policy Directive). 2011. "Presidential policy directive 8 and the national preparedness system: Background and issues for congress." Accessed August 13, 2019. https://fas.org/sgp/crs/homesec/R42073.pdf.

PPD. 2013. "Critical infrastructure security and resilience." PPD-21. Accessed May 12, 2019. https://obamawhitehouse.archives.gov/the-press-office/2013/02/12/presidential-policy-directive-critical-infrastructure-security and-resil

Rashedi, R., and T. Hegazy. 2016. "Holistic analysis of infrastructure deterioration and rehabilitation using system dynamics." *J. Infrastruct. Syst.* 22 (1): 04015016.

Rasoulkhani, K., B. Logasa, M. Presa Reyes, and A. Mostafavi. 2018. "Understanding fundamental phenomena affecting the water conservation technology adoption of residential consumers using agent-based modeling." *Water* 10 (8): 993.

Rasoulkhani, K., and A. Mostafavi. 2018. "Resilience as an emergent property of human-infrastructure dynamics: A multi-agent simulation model for characterizing regime shifts and tipping point behaviors in infrastructure systems." *PLoS One* 13 (11): e0207674.

Rasoulkhani, K., A. Mostafavi, J. Cole, and S. Sharvelle. 2019. "Resilience-based infrastructure planning and asset management: Study of dual and singular water distribution infrastructure performance using a simulation approach." *Sustainable Cities Soc.* 48: 101577.

Rechtin, E. 1991. *Systems architecting: Creating and building complex systems.* Englewood Cliffs, NJ: Prentice Hall.

Reed, D. A., K. C. Kapur, and R. D. Christie. 2009. "Methodology for assessing the resilience of networked infrastructure." *IEEE Syst. J.* 3 (2): 174–180.

Reed, D. A., S. Wang, K. C. Kapur, and C. Zheng. 2015. "Systems-based approach to interdependent electric power delivery and telecommunications infrastructure resilience subject to weather-related hazards." *J. Struct. Eng.* 142 (8): C4015011.

Rehan, R., M. A. Knight, C. T. Haas, and A. J. A. Unger. 2011. "Application of system dynamics for developing financially self-sustaining management policies for water and wastewater systems." *Water Res.* 45 (16): 4737–4750.

Rose, A. 2007. "Economic resilience to natural and man-made disasters: Multidisciplinary origins and contextual dimensions." *Environ. Hazards* 7: 383–398.

Saadat, Y., B. M. Ayyub, Y. J. Zhang, D. M. Zhang, and H. W. Huang. 2019. "Resilience of metrorail networks: Quantification with Washington D.C. as a case study." *ASCE-ASME J. Risk Uncertainty Eng. Syst. Part B: Mech. Eng.* 5 (4): 041011.

Saadat, Y., B. M. Ayyub, Y. Zhang, D. Zhang, and H. Huang. 2020. "Resilience-based strategies for topology enhancement and recovery of metrorail transit networks." *ASCE-ASME J. Risk Uncertainty Eng. Syst. Part A: Civ. Eng.* 6 (2): 04020017.

Sanford Bernhardt, K. L., and S. McNeil. 2008. "Agent-based modeling: Approach for improving infrastructure management." *J. Infrastruct. Syst.* 14 (3): 253–261.

Sharma, N., and P. Gardoni. 2019. "Modeling the time-varying performance of electrical infrastructure during post disaster recovery using tensors." In *Handbook of sustainable and resilient infrastructure*, edited by P. Gardoni, 259–276. Abington, UK: Routledge.

Sharma, N., A. Tabandeh, and P. Gardoni. 2018. "Resilience analysis: A mathematical formulation to model resilience of engineering systems." *Sustainable Resilient Infrastruct.* 3 (2): 49–67.

Sharma, N., A. Tabandeh, and P. Gardoni. 2019a. "Regional resilience analysis: A multi-scale formulation to model the recovery of interdependent infrastructure." In *Handbook of sustainable and resilient infrastructure*, edited by P. Gardoni, 521–544. Abington, UK: Routledge.

Sharma, N., A. Tabandeh, and P. Gardoni. 2019b. "Recovery optimization of interdependent infrastructure: A multi-scale approach." In *Proc., 13th Int. Conf. on Applications of Statistics and Probability in Civil Engineering.*

Shinozuka, M., A. Rose, and R. T. Eguchi. 1998. *Engineering and socioeconomic impacts of earthquakes.* Monograph 98-MN02. Buffalo, NY:

Multidisciplinary Center for Earthquake Engineering Research (MCEER).

Steelman, J., J. Song, and F. Jerome. 2007. *Integrated data flow and risk aggregation for consequence-based risk management of seismic regional losses*. Technical Rep. Urbana, IL: Mid-America Earthquake Center, University of Illinois at Urbana-Champaign.

Tabandeh, A., P. Gardoni, C. Murphy, and N. Myers. 2019. "Societal risk and resilience analysis: Dynamic Bayesian network formulation of a capability approach." *ASCE-ASME J. Risk Uncertainty Eng. Syst. Part A: Civ. Eng.* 5 (1): 04018046.

Tierney, K., and M. Bruneau. 2007. "Conceptualized and measuring resilience." *TR News* 250: 14–17. Accessed April 8, 2015. http://onlinepubs.trb.org/onlinepubs/trnews/trnews250_p14-17.pdf.

Vugrin, E. D., M. J. Baca, M. D. Mitchell, and K. L. Stamber. 2014a. "Evaluating the effect of resource constraints on resilience of bulk power system with an electric power restoration model." *Int. J. Syst. Syst. Eng.* 5 (1): 68–91.

Vugrin, E. D., M. A. Turnquist, and N. J. K. Brown. 2010. *Optimal recovery sequencing for critical infrastructure resilience assessment*. Technical Rep. SAND2010-6237. Albuquerque, NM: Sandia National Laboratories.

Vugrin, E. D., M. A. Turnquist, and N. J. K. Brown. 2014b. "Optimal recovery sequencing for enhanced resilience and service restoration in transportation networks." *Int. J. Crit. Infrastruct.* 10 (3/4): 218–246.

Vugrin, E. D., S. J. Verzi, P. D. Finley, M. A. Turnquist, A. R. Griffin, K. A. Ricci, and T. Wyte-Lake. 2015. "Modeling evacuation of a hospital without electric power." *Prehospital Disaster Med.* 30 (3): 279–287.

Vugrin, E. D., D. E. Warren, and M. A. Ehlen. 2011. "A resilience assessment framework for infrastructure and economic systems: Quantitative and qualitative resilience analysis of petrochemical supply chains to a hurricane." *Process Saf. Prog.* 30 (3): 280–290.

Washington DC Metro Fiscal Year (FY) Budget. 2018. Accessed October 31, 2018. https://www.wmata.com/about/records/public_docs/upload/Metro_FY2018_Proposed_Budget_15Dec16_v4.pdf.

Zhu, J., and A. Mostafavi. 2014. "Towards a new paradigm for management of complex engineering projects: A system-of-systems framework." In *Proc., IEEE Int. Systems Conf.* 213–219, 2014. Ottawa, ON.

Zhu, J., and A. Mostafavi. 2018. "Performance assessment in complex engineering projects using a system-of-systems framework." *IEEE Syst. J.* 12 (1): 262–273.

CHAPTER 4

RESILIENCE ECONOMICS AND RISK MANAGEMENT

4.1 PLANNING HORIZON AND DISCOUNT RATES

The planning horizon defines the length of the study period, such as the design life of an infrastructure system. The study period should be sufficiently long enough to capture all impacts (e.g., risk reduction) generated from those intervention expenditures (costs) under analysis to accurately evaluate the returns of those investments. Study periods that are insufficiently short risk undervaluing investments in resilience. Study periods that are excessively long risk overvaluing investments. The planning horizon should be coordinated with the discount rate selected. For example, accounting for a changing climate in designing critical infrastructure with an expected long-utilization time might require a long planning horizon of at least 100 years and a small discount rate; the discount rate in this case does not shift the risk to future generations.

The discount rate captures the time value of money. Discounting is used to normalize future values into present values, facilitating comparisons of cash flows occurring at different times across the study period. High discount rates shift the relative importance of cash flows earlier in the study period, meaning those costs and losses that occur earlier in the study period have a disproportionate effect on the results than do values that occur later in the study period. By comparison, low discount rates effectively allow values farther in the future to have a greater influence on the results.

The discount rate is sometimes referred to as a *hurdle rate*, interest rate, cutoff rate, benchmark, or the cost of capital (Defusco et al. 2001, Brealey and Myers 2000). Some firms have a selected discount rate for all project analyses; however, risk can often be a determinant for selecting a rate, similar to how banks have high interest rates for loaning money to high-risk individuals. Selecting a discount rate is often a challenge. In general, it is the minimum rate of return that one would need to engage in a particular investment and is greater than or equal to the return on other readily available investments such as stocks and bonds.

A key component for evaluating an investment decision is the time value of money, that is, the relationship between cash flows occurring at different time periods. Receiving \$100 today is, in general, preferred to receiving the same amount a year later. To account for these different values, future cash flows are *discounted* to equate them to values received today (Defusco et al. 2015). This is calculated by dividing future cash flows with an interest rate or discount rate (Thomas 2017)

$$\mathrm{PV}_1 = \frac{\mathrm{CF}_1}{1+r} \tag{4-1}$$

where
PV_1 = Present value of future cash flow after 1 year,
CF_1 = Cash flow after 1 year, and
r = Discount rate, which is, typically, between 0 and 1.

Selection of a discount rate is an important decision because it can profoundly impact the analysis results. When the discount rate is set to account for the opportunity cost of capital, it provides a mechanism to evaluate the return from investing in resilience compared with other investment alternatives. When a real discount rate is used, as opposed to a nominal rate, both opportunity cost and inflation have been accounted for.

4.2 STANDARD APPROACHES FOR EVALUATING INVESTMENTS

A number of standard economic approaches for evaluating and comparing investments are presented in this section. They include benefit–cost analysis (BCA) or net present value (NPV), life-cycle cost analysis (LCCA), savings-to-investment ratio (SIR), internal rate of return (IRR), decision trees and real options, and sensitivity analysis with Monte Carlo techniques. For each approach presented, its definition, use, method (formula), interpretation, and an example are provided. It should be noted that behavioral economics, expected utility, risk aversion, prospect theory, and so on that are not covered in this chapter reinforce the point to engineers that humans do not always make rational decisions (Ayyub 2014).

Although there are a number of approaches commonly used for investment analysis, some are more prevalent than others. As seen in Table 4-1, the IRR and net present value (NPV) are more frequently used

Table 4-1. Survey Response to "How Frequently Does Your Firm Use the Following Techniques When Deciding Which Projects or Acquisitions to Pursue."

Attribute	% Always or Almost Always	Average Response[a]	Average Response by Firm Size[a]	
			Small	Large
Internal rate of return	75.61	3.09	2.87	3.41[d]
Net present value	74.93	3.08	2.83	3.42[d]
Payback period	56.74	2.53	2.72	2.25[d]
Hurdle rate	56.94	2.48	2.13	2.95[d]
Sensitivity analysis	51.54	2.31	2.13	2.56[d]
Earnings multiple approach	38.92	1.89	1.79	2.01[b]
Discounted payback period	29.45	1.56	1.58	1.55
We incorporate the *real options* of a project when evaluating it	26.59	1.47	1.4	1.57
Accounting rate of return	20.29	1.34	1.41	1.25
Value-at-risk or other simulation	13.66	0.95	0.76	1.22[d]
Adjusted present value	10.78	0.85	0.93	0.72
Profitability index	11.87	0.83	0.88	0.75

[a]Respondents were asked on a scale from 0 (never use) to 4 (always use).

[b]Statistically different from small at the 1% level.

[c]Statistically different from small at the 5% level (none of the cases).

[d]Statistically different from small at the 10% level.

Source: Adapted from Graham and Harvey (2001).

than many other approaches. The data in this table are based on a survey of 392 chief financial officers (CFOs) about the cost of capital, capital budgeting, and capital structure. Surveys were sent to CFOs for firms listed in the Fortune 500 rankings with approximately 40% of them being manufacturers and 15% being financial. The survey asked on a scale from

0 to 4, "how frequently does your firm use the following techniques when deciding which projects or acquisitions to pursue" with 0 representing *never use it* and 4 meaning *always use it*. Responses vary by firm size, as seen in the table.

The different methods for investment analysis are applicable to certain decision types and have some limitations. As seen in Table 4-2, all can be used for an accept/reject decision for an investment as well as design and size, whereas others can be used to prioritize or rank investments. Examples of the various decision types are provided in Table 4-3, and limitations of each approach are discussed in Table 4-4. Some approaches require that the comparison is made over the same study period, but, in cases when the study periods differ, the stream of costs/benefits of the original investment is assumed to repeat (some methods do not consider information about the duration of a project).

4.2.1 Benefit–Cost Analysis Using Net Present Value

This section introduces BCA based on NPV by providing a definition, suggested use, and the underlying method.

Table 4-2. Application of Methods for Investment Analysis.

Attribute	Net Present Value	Life-Cycle Cost	Savings-to-Investment Ratio	Internal Rate of Return	Monte Carlo Analysis	Decision Trees
Accept/reject	X	X	X	X	—	—
Design	X	X	X*	X*	—	—
Size	X	X	X*	X*	—	—
Priority or ranking	X		X	X	—	—
Uncertainty and potential outcomes	—	—	—	—	X	X

*Appropriate when incremental discounted costs and benefits are considered (i.e., the difference in costs/benefits between two investments). To decide between more than two options, pairwise comparisons are necessary.

Source: Thomas (2017).

Table 4-3. Examples of Decision Types.

Accept/reject	• Is an investment in a backup generator cost effective? • Is elevating HVAC control systems in emergency response facilities cost effective? • Is it cost effective to retro fit for earthquake resistance?
Design	• Which bridge design is the most cost effective? • Which seawall design is the most cost effective? • Which flood control system is the most cost effective? • Which roofing system is the most cost effective?
Size	• What size of bridge is the most cost effective? • What height of seawall is the most cost effective?
Priority or ranking	• Is it more cost effective to invest in wider roads or a flood control system? • We have five proposed investments but can only afford a selection of them. Which investments do we choose?
Uncertainty and potential outcomes	• We are considering an investment in a flood management system. We need to consider the potential for varying flood levels • We are considering an investment to move electrical lines underground. We need to consider the uncertainty in the risk of flooding compared to that of high winds. • We are considering an investment in wastewater management. We need to consider the uncertainty in flooding.

Source: Thomas (2017).

Definition: BCA is a method for analyzing an investment that incorporates investment expenditures, costs, savings, income, and the discount rate.

Use: This technique is applicable to accept/reject, design, size, and priority/ranking decisions. Alternatives must be compared over the same study period.

Method: BCA primarily relies on the NPV, which is the difference between the present value of all cash inflows and the present value of all cash outflows over the period of the investment (Defusco et al. 2001, 2015; Budnick 1998). It is calculated by adjusting each monetary cost and benefit associated with an investment to a common time, in this instance, time zero. The adjustment accounts for the time value of money, as described, along with the decreased purchase power of money due to inflation. Inflows (benefits) are summed together with the outflows (costs) being subtracted:

Table 4-4. Limitations and Considerations of Methods for Investment Analysis.

Method	Limitation
Net present value	Alternatives must be compared over the same study period
Life-cycle cost	Alternatives must be compared over the same study period. A single LCC does not indicate the merit of an investment, as it must be compared with an alternative
Savings-to-investment ratio	If reinvestment earnings are not equal the discount rate, then there may be inconsistent results. Alternatives must be compared over the same study period
Internal rate of return	In some instances, inconsistent results may arise. This calculation does not reveal information about the size or duration of a project. Alternatives must be compared over the same study period or it must be assumed that assets can be expected to repeat the costs/benefits of the original investment

Source: Thomas (2017).

$$\mathrm{NPV} = -C_0 + I_0 + \frac{-C_1}{(1+r)} + \frac{B_1}{(1+r)} + \frac{-C_2}{(1+r)^2} + \frac{B_2}{(1+r)^2} + \cdots + \frac{-C_T}{(1+r)^T} + \frac{B_T}{(1+r)^T} \tag{4-2}$$

where
B_t = Total benefits, which includes cash inflow and savings, in time period t,
C_t = Total cost in time period t,
r = Discount rate,
t = Time period, which is typically measured in years, and
T = Total number of years (i.e., planning horizon or study period length).

Equation (4-2) can be written in a concise manner as follows:

$$\mathrm{NPV} = \sum_{t=0}^{T} \frac{(B_t - C_t)}{(1+r)^t} \tag{4-3}$$

For each time, net cash inflows are divided by 1 plus a selected discount rate raised to the power of the time period, t. Either a nominal or real discount rate can be used, which is determined by whether it is a current or constant dollar analysis. Costs and benefits are not adjusted for inflation in a current dollar analysis. In a constant dollar analysis, costs and benefits are adjusted for inflation.

Interpretation: For accept/reject decisions, an investment can be accepted when the NPV is positive or equal to zero and rejected when it is negative. For design, size, and priority ranking, an investment is ranked higher when it has a higher NPV. In the case of priority/ranking, additional measures, such as saving-to-investment ratio or IRR, should also be considered, because the NPV does not take into account the magnitude of the investment.

Example: The city of Springfield is considering an investment into elevating various electrical equipment for their four school buildings (School A, School B, School C, and School D) to protect against flooding at a cost of $30k for School A, $20k each for Schools B and C, and $50k for School D, as seen in Table 4-5. There is a 2% annual probability of a flood that would damage the electrical equipment at School A with a cost of $400k to repair/replace the equipment. School B has a 1% probability with the same repair costs. Schools C and D have a 0.2% chance of a flood that damages the electrical equipment with a cost of $400k and $500k to repair it. The city is required to use a 5% real discount rate and a 30 year study period, because the city might replace the schools after 30 years. The annual expected loss is calculated as the probability multiplied by the repair cost and is shown in the last column of Table 4-5. Table 4-6 presents the annual present values for Springfield's investment—note that other metrics are also presented, which are discussed later. Columns C, F, I, and L present values of averted losses

Table 4-5. Example of Net Present Value, Data.

School	Investment ($000s)	Probability of Flood (%)	Repair Cost ($000s)	Annual Expected Loss ($000s), Current Dollars
School A	30	2.0	400	8
School B	20	1.0	400	4
School C	20	0.2	400	0.8
School D	50	0.2	500	1

Source: Thomas (2017).

Table 4-6. Annual Present Value for Springfield Investment ($000s) Using 5% Real Discount Rate.

	A	B	C	D	E	F	G	H	I	J	K	L	M
	School A			School B			School C			School D			
Year	Costs – no retrofit	Costs – retrofit	Difference (A – B)	Costs – no retrofit	Costs – retrofit	Difference (D – E)	Costs – no retrofit	Costs – retrofit	Difference (G – H)	Costs – no retrofit	Costs – retrofit	Difference (J – K)	Annual Total for NPV M = C + F + I + L
0	0	30	−30	0	30	−30	0	20	−20	0	50	−50	−130
1	7.6	0.0	7.6	3.8	0.0	3.8	0.8	0.0	0.8	1.0	0.0	1.0	13.1
2	7.3	0.0	7.3	3.6	0.0	3.6	0.7	0.0	0.7	0.9	0.0	0.9	12.5
3	6.9	0.0	6.9	3.5	0.0	3.5	0.7	0.0	0.7	0.9	0.0	0.9	11.9
4	6.6	0.0	6.6	3.3	0.0	3.3	0.7	0.0	0.7	0.8	0.0	0.8	11.4
5	6.3	0.0	6.3	3.1	0.0	3.1	0.6	0.0	0.6	0.8	0.0	0.8	10.8
6	6.0	0.0	6.0	3.0	0.0	3.0	0.6	0.0	0.6	0.7	0.0	0.7	10.3
7	5.7	0.0	5.7	2.8	0.0	2.8	0.6	0.0	0.6	0.7	0.0	0.7	9.8
8	5.4	0.0	5.4	2.7	0.0	2.7	0.5	0.0	0.5	0.7	0.0	0.7	9.3

9	5.2	0.0	5.2	2.6	0.0	2.6	0.5	0.0	0.5	0.6	0.0	0.6	8.9
10	4.9	0.0	4.9	2.5	0.0	2.5	0.5	0.0	0.5	0.6	0.0	0.6	8.5
11	4.7	0.0	4.7	2.3	0.0	2.3	0.5	0.0	0.5	0.6	0.0	0.6	8.1
12	4.5	0.0	4.5	2.2	0.0	2.2	0.4	0.0	0.4	0.6	0.0	0.6	7.7
13	4.2	0.0	4.2	2.1	0.0	2.1	0.4	0.0	0.4	0.5	0.0	0.5	7.3
14	4.0	0.0	4.0	2.0	0.0	2.0	0.4	0.0	0.4	0.5	0.0	0.5	7.0
15	3.8	0.0	3.8	1.9	0.0	1.9	0.4	0.0	0.4	0.5	0.0	0.5	6.6
16	3.7	0.0	3.7	1.8	0.0	1.8	0.4	0.0	0.4	0.5	0.0	0.5	6.3
17	3.5	0.0	3.5	1.7	0.0	1.7	0.3	0.0	0.3	0.4	0.0	0.4	6.0
18	3.3	0.0	3.3	1.7	0.0	1.7	0.3	0.0	0.3	0.4	0.0	0.4	5.7
19	3.2	0.0	3.2	1.6	0.0	1.6	0.3	0.0	0.3	0.4	0.0	0.4	5.5
20	3.0	0.0	3.0	1.5	0.0	1.5	0.3	0.0	0.3	0.4	0.0	0.4	5.2
21	2.9	0.0	2.9	1.4	0.0	1.4	0.3	0.0	0.3	0.4	0.0	0.4	5.0
22	2.7	0.0	2.7	1.4	0.0	1.4	0.3	0.0	0.3	0.3	0.0	0.3	4.7
23	2.6	0.0	2.6	1.3	0.0	1.3	0.3	0.0	0.3	0.3	0.0	0.3	4.5
24	2.5	0.0	2.5	1.2	0.0	1.2	0.2	0.0	0.2	0.3	0.0	0.3	4.3
25	2.4	0.0	2.4	1.2	0.0	1.2	0.2	0.0	0.2	0.3	0.0	0.3	4.1

(Continued)

Table 4-6. (*Continued*)

	A	B	C	D	E	F	G	H	I	J	K	L	M
	School A			School B			School C			School D			
Year	Costs – no retrofit	Costs – retrofit	Difference (A – B)	Costs – no retrofit	Costs – retrofit	Difference (D – E)	Costs – no retrofit	Costs – retrofit	Difference (G – H)	Costs – no retrofit	Costs – retrofit	Difference (J – K)	Annual Total for NPV M = C + F + I + L
26	2.2	0.0	2.2	1.1	0.0	1.1	0.2	0.0	0.2	0.3	0.0	0.3	3.9
27	2.1	0.0	2.1	1.1	0.0	1.1	0.2	0.0	0.2	0.3	0.0	0.3	3.7
28	2.0	0.0	2.0	1.0	0.0	1.0	0.2	0.0	0.2	0.3	0.0	0.3	3.5
29	1.9	0.0	1.9	1.0	0.0	1.0	0.2	0.0	0.2	0.2	0.0	0.2	3.4
30	1.9	0.0	1.9	0.9	0.0	0.9	0.2	0.0	0.2	0.2	0.0	0.2	3.2
NPV													82.1
LCC	123.0	30.0		61.5	30.0		12.3	20.0		15.4	50.0		
SIR			4.10			2.05			0.61			0.31	
IRR			0.27			0.13			0.01			<0	

Source: Thomas (2017).

Note: NPV = net present value; LCC = life-cycle cost; SIR = savings-to-investment ratio; and IRR = internal rate of return.

minus investment costs. The NPV for this investment can be calculated as follows:

$$\text{NPV for retrofitting} = \sum_{y=0}^{30} M_y \tag{4-4}$$

where M_y is the value in Column M of Table 4-6 corresponding to year y.

The sum of the present values is \$82.1k (see the bottom of Column M in Table 4-6), which is greater than zero; therefore, the investment is economical.

4.2.2 Life-Cycle Cost Analysis

This section introduces life-cycle cost analysis (LCCA) by providing a definition, suggested use, and the underlying method.

Definition: Life-cycle cost (LCC) measures all relevant costs of a potential investment in present-value terms. It is a variant of NPV except that it does not consider benefits other than those realized as lower costs (savings). The focus is cost effectiveness.

Use: This technique is applicable to accept/reject, design, and size decisions. It is particularly useful for examining whether initial higher costs are justified by future lower costs. Alternatives must be compared over the same study period. Unlike NPV, a single LCC does not indicate the merit of an investment, as it must be compared with an alternative.

Method: LCC is the sum of present-value costs and can be calculated as follows:

$$\text{LCC} = \sum_{t=0}^{T} \frac{(C_t)}{(1+r)^t} \tag{4-5}$$

Interpretation: If an investment has a lower LCC than another investment, it is considered to be more cost effective.

Example: The city of Springfield is reconsidering their investment in elevating their electrical equipment, as discussed in Section 4.2.1. Now, they are considering whether all schools should have their systems elevated or only a selection of them. This is a question of the size of the investment. The LCC for School A with no retrofit is calculated as follows:

$$\text{LCC School A no retrofit} = \sum_{y=0}^{30} A_y \tag{4-6}$$

where A_y is the value in Column A of Table 4-6 corresponding to year y.

The other LCCs are the individual sum of Columns B, D, E, G, H, J, and K with each representing the corresponding school and designated retrofit or lack thereof. The LCCs are shown at the bottom of Table 4-6.

Using the LCC, it is cost effective for School A and School B to elevate their systems, as the LCC to do so is less than not doing it (i.e., comparing the LCC at the bottom of Columns A and B along with comparing that of Columns G and H in Table 4-6). However, it is not cost effective for School C or School D, as the LCC in Columns G and J is less than that in Columns H and K.

4.2.3 Savings-to-Investment Ratio

This section introduces the SIR method by providing a definition, suggested use, and the underlying method.

Definition: SIR is discounted savings divided by discounted investment costs.

Use: SIR is appropriate for accept/reject, design, size, and priority/ranking decisions. If reinvestment earnings are not equal to the discount rate, then there may be inconsistent results. Alternatives must be compared over the same study period.

Method: SIR is calculated as follows:

$$\mathrm{SIR} = \frac{\sum_{t=0}^{T} S_t/(1+r)^t}{\sum_{t=0}^{T} I_t/(1+r)^t} \tag{4-7}$$

where S_t is the cost savings in period t.

Interpretation: Investments with higher SIRs are more economical than those with lower ratios. Those that are above 1 are considered cost effective, whereas those below 1 are not cost effective.

Example: The city of Springfield is again reconsidering its investment in elevating various electrical systems (Section 4.2.1). In addition to considering the LCC, it examines the savings-to-investment ratio. The ratio is calculated as follows:

$$\text{SIR School A} = \frac{\sum_{y=1}^{30} C_y}{C_1} \tag{4-8}$$

where C_y is the value in Column C of Table 4-6 corresponding to year y.

For subsequent values, Columns F, I, and L are used for their corresponding schools. The ratios are shown at the bottom of Table 4-6.

For Schools A and B, the ratio is greater than 1, whereas for Schools C and D, the value is less than 1; therefore, the investments in Schools A and B are cost effective, whereas those in C and D are not.

4.2.4 Internal Rate of Return

This section introduces the IRR method by providing a definition, suggested use, and the underlying method.

Definition: The IRR is the discount rate that results in the NPV equaling zero.

Use: The IRR is applicable in examining accept/reject, design, size, and priority/ranking decisions. In some instances, inconsistent results may arise. This calculation does not reveal information about the size or duration of a project. Alternatives must be compared over the same study period or it must be assumed that assets can be expected to repeat the costs/benefits of the original investment.

Method: Owing to the nature of this calculation, individuals often use software or trial and error to identify the IRR (i.e., select varying discount rates to identify the value where the NPV equals zero).

Interpretation: A higher IRR is more economic than a lower 1. An IRR that is greater than the selected discount rate is a cost-effective investment, whereas lower values are not.

Example: The decision makers in Springfield, from the example in Section 4.2.1, realize they can only retrofit one school. To determine which school is the most cost effective, they examine the IRR. As seen at the bottom of Table 4-6, one can observe that School A has the largest rate; thus, this is the most economic choice. Schools C and D are below the city's discount rate of 5%; thus, they are not economical. School B has a lower rate than School A; therefore, it is not as cost effective as School A.

4.2.5 Decision Trees and Real Options

This section introduces decision trees and real options by providing a definition, suggested use, and the underlying method.

Definition: Real options are choices an organization makes at various time points in an investment. This is often traced using a decision tree where decisions are drawn out in sequence. For instance, if a new product is not successful, then there is the option to abandon it. There is also an option to expand it if it is successful.

Use: Real options are used when there are choices that a decision maker can choose at various time points in an investment. Options pricing theory is an advanced topic, which is not completely covered in this document. For more information, one might refer to Copeland and Antikarov (2003) or Brealey and Myers (2000).

Method: Real options pricing is not fully discussed here; however, it can be described in a decision tree. There are, typically, three types of nodes in a decision tree: decision nodes represented by squares, chance nodes represented by circles, and end nodes represented by triangles. Nodes are connected in a treelike manner. Real options suggest that the total value of a project is the NPV plus the value of options

$$TPV = NPV + VO \tag{4-9}$$

where TPV is the total project value, and VO is the value of options.

Interpretation: Real options and decision trees can utilize different metrics; thus, the interpretation depends on the metric selected. In general, the decision tree aids in identifying which paths/decisions one might take in the future.

Example: Recall the example in Section 4.2.1, where the city of Springfield is considering elevating various electrical equipment. The city decides to invest in raising the electrical equipment for School A. However, there is a 30% probability that funding will be available to retrofit School B in the fifth year of the study period. Although retrofitting School B was economical in Year 0, it may not be economical in the fifth year of the study period. To examine this option, the city uses real options with a decision tree. Following along from left to right in the decision tree in Figure 4-1, the initial investment is $30k for retrofitting School A. Then, there is the

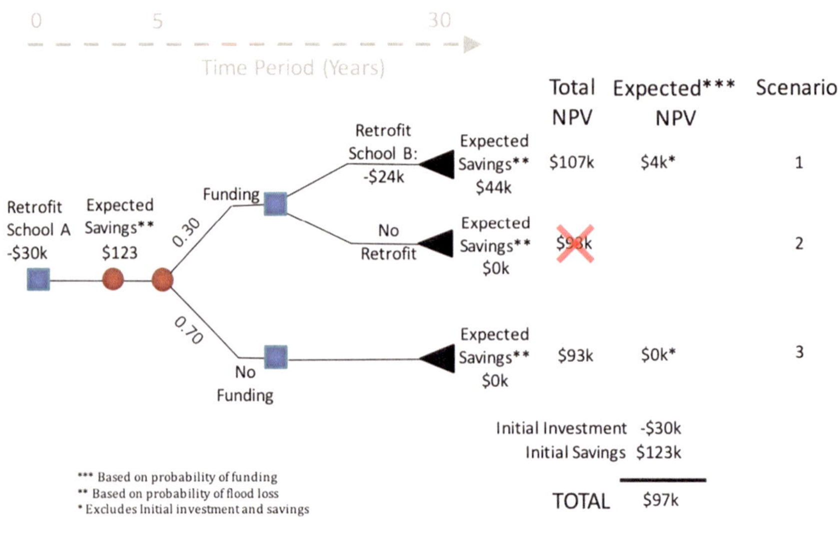

Figure 4-1. Decision tree example, discounted values.
Source: Thomas (2017).

expected savings of $123k from retrofitting School A, which is the sum of Years 1 through 30 in Column C in Table 4-6. There is a 0.30 chance that there is funding to retrofit School B with the cost being another $30k, which is discounted to $24k. With the retrofit, there are savings from expected flooding loss, which is the sum of Years 6 through 30 in Column F. This is Scenario 1, as labeled on the right side of Figure 4-1. Without the retrofit (see Scenarios 2 and 3), there is no additional savings. The NPV for each scenario is calculated as follows:

$$\text{Total NPV Scenario 1} = -30 + \sum_{y=1}^{30} C_y + \left(\frac{-30}{1.05^5}\right) + \sum_{y=6}^{30} F_y = 107 \quad (4\text{-}10)$$

$$\text{Total NPV Scenario 2} = -30 + \sum_{y=1}^{30} C_y = 93 \quad (4\text{-}11)$$

$$\text{Total NPV Scenario 3} = -30 + \sum_{y=1}^{30} C_y = 93 \quad (4\text{-}12)$$

where F_y is the value in Column F of Table 4-6 corresponding to year y.

The NPV for Scenario 1 exceeds that of Scenario 2; thus, we eliminate Scenario 2, as it is not economical, and it would not be chosen. The expected value for the investment can be estimated as the sum of the initial investment and savings plus the expected value of Scenarios 1 and 3, which are represented as follows:

$$\text{Expected NPV Scenario 1} = 0.30 * \left[\left(\frac{-30}{1.05^5}\right) + \sum_{y=6}^{30} F_y\right] = 4 \quad (4\text{-}13)$$

$$\text{Expected NPV Scenario 3} = 0.70 * 0 = 0 \quad (4\text{-}14)$$

After adding the initial investment and savings, the expected value with options is $97k. Without options, the NPV considering School A and School B would be the NPV of Scenario 3, where only School A is retrofitted or $93k.

4.2.6 Sensitivity Analysis with Monte Carlo Techniques

This section introduces sensitivity analysis with Monte Carlo techniques by providing a definition, suggested use, and the underlying method.

Definition: Monte Carlo analysis uses sampling techniques that sample from user-specified distributions for a selection of variables. This process then creates a distribution of potential outcomes. This technique is described by Ayyub and McCuen (2011).

Use: Monte Carlo analysis is applicable when there are variables within an investment that have uncertain values.

Method: This technique is, typically, implemented using a software package such as NIST's free Monte Carlo Tool (Thomas 2019). Specification involves defining which variables are to be simulated, the distribution of each of these variables, and the number of iterations performed. The software then randomly samples from the probabilities for each input variable of interest. Three common distributions that are used include triangular, normal, and uniform.

Interpretation: Monte Carlo analysis can utilize different metrics; therefore, the interpretation depends on the metric selected. However, the output is frequently placed into a cumulative probability graph to examine the probability of different outcomes.

Example: Recall the example in Section 4.2.1, where the city of Springfield is considering elevating various electrical equipment. In Section 4.2.4, the city used the IRR to choose one school to retrofit; however, it turns out that the nearby water management system has been renovated and the effects are not well known for their area. Thus, the probability of flooding for School A and School B is uncertain.

It is determined that for School A, the annual probability is most likely 2.0%, but it could be as low as 0.2% and as high as 2.5%. For School B, the probability could be as low as 0.8% and as high as 3.0%. The city is also uncertain with the discount rate that they have chosen. To incorporate these uncertainties into their analysis, the city uses Monte Carlo analysis. For the probability of flood at School A, a triangular distribution is used with 0.2% as low, 2.5% as high, and 2.0% as the most likely. For the probability of flood at School B, a triangular distribution is used with 0.8% as low, 1.0% as high, and 3% as the most likely. The discount rate was varied between 3% and 7% using a uniform distribution. A simulation of 5k trials was used and implemented using NIST's free Monte Carlo Tool with the SIR as the metric to compare investments.

As illustrated in Figure 4-2, 2% of School A's trials were not economical (i.e., the SIR was less than 1). Figure 4-3 shows the cumulative probability graph of the SIR for School A divided by that of School B from the Monte Carlo analysis. As illustrated in the table, for 2,539 of the 5,000 simulations or 50.8%, the ratio is less than 1, where the SIR for School B exceeds that of School A. Therefore, given the selected distributions, School A has a slightly higher probability of being more cost effective; however, it is close enough that either option could be selected.

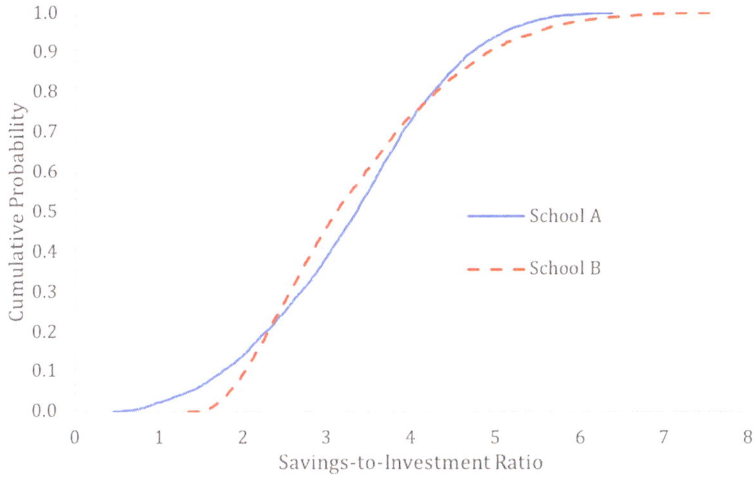

Figure 4-2. Cumulative probability of the savings-to-investment ratio from Monte Carlo analysis (each school).

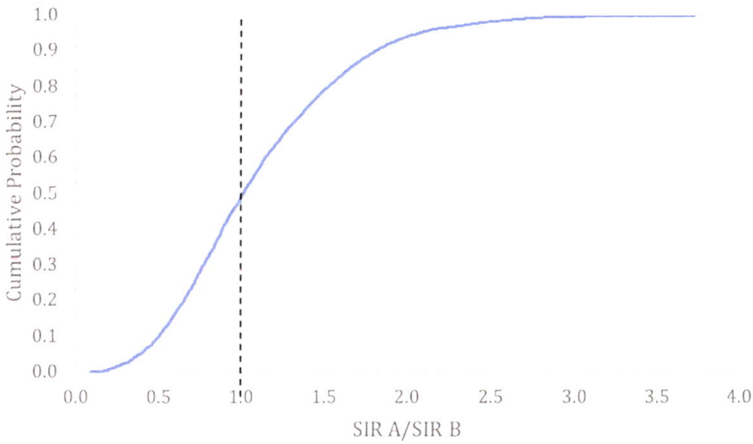

Figure 4-3. Cumulative probability of the savings-to-investment ratio from Monte Carlo analysis (Ratio A divided by Ratio B).

4.3 COST CONSIDERATIONS

Costs, which are expenditures made to improve resilience, can be organized into phases: preparedness, response, and recovery. Although these phases are defined by the timing of the activities relative to a hazard event, the unit prices of individual actions or investments may vary depending on the severity and duration of the hazard event. Scarcity can occur through capital destruction or supply chain disruption during and after an event and impact prices. Depending on the availability of substitutes, resource costs may be affected and also difficult to forecast ahead of time.

Costs can also be categorized based on timing (initial versus noninitial), type of impact (direct versus indirect), and whether they produce an externality (i.e., affect a third party). Indirect costs and externalities are often the most difficult to estimate, particularly when compared with initial investment costs. Costs can be adapted to include price escalation rates when future prices are expected to increase in real (inflation adjusted) terms.

4.4 EXPECTED LOSS CONSIDERATIONS

Losses represent the dollar value of damage resulting from disruptions and disasters. Expected losses are those losses that are estimated as those likely to occur in the future given current and expected investments into resilience. Expected losses can be represented mathematically as the probability of a disruption or disaster multiplied by its consequence (in dollars) and thus are sensitive to the timing and severity of the event.

Models are often required to estimate expected losses. Theoretical approaches can often be intractable or fail to provide a closed-form solution. Several other approaches exist. These can be grouped into (1) empirically based or (2) regional economic model (simulation-based) methods. A brief introduction follows.

4.4.1 Empirical Methods

Econometric and statistical analyses based on real-world data are useful in establishing correlations and, potentially, causal effects related to losses from natural disasters (Fung and Helgeson 2017). In particular, causal inference (or policy evaluation in the economics literature) can be used to evaluate the direct or indirect impacts of a natural disaster by estimating economic outcomes relative to the counterfactual in which no disaster occurs.

Cavallo et al. (2013) use the method of synthetic control to estimate the causal impact of natural disasters on short- (up to 5 years) and long-run (beyond 5 years) economic growth. After controlling for political shocks (e.g., a political revolution), they find that there is no significant impact on either short- or long-run growth. In terms of policy analysis, Fu and Gregory (2019) exploit a discontinuity in the policy for awarding Post-Katrina rebuilding grants in New Orleans, Louisiana, to show that the grant program generates positive externalities to communities in which rebuilding occurs. Finally, Cavallo and Noy (2009) provide a literature review of the economics of natural disasters, including a review of available disaster data, estimates of direct effects, and estimates of short- and long-run indirect effects.

Beyond causal inference, data mining and machine learning methods could be used to infer patterns from data. In turn, these may shed light on what makes some communities more resilient than others, and in turn, aid in comparing alternative approaches to enhancing resilience.

4.4.2 Input–Output and Computable General Equilibrium Methods

Input–output (IO) and computable general equilibrium (CGE) models are a class of methods that model economic interdependencies within a regional economy. Such macroeconomic models are useful for identifying *higher order effects* of natural disasters (Okuyama and Santos 2014). CGE models simulate a complete market economy in which prices and quantities supplied and demanded adjust to clear all markets. In contrast, IO models do not respond to price changes, but they are simpler and, therefore, easier to work with than CGE models.

Figure 4-4 provides a general conceptualization of a *Social Accounting Matrix* (SAM), a major component of a CGE model. The *SAM Region* defines the study area. Households and producers are linked through product and capital markets through savings and sales. Government is linked through households and producers through taxes and transfers. Flows to and from areas outside the SAM region (*Rest of World*) are allowed.

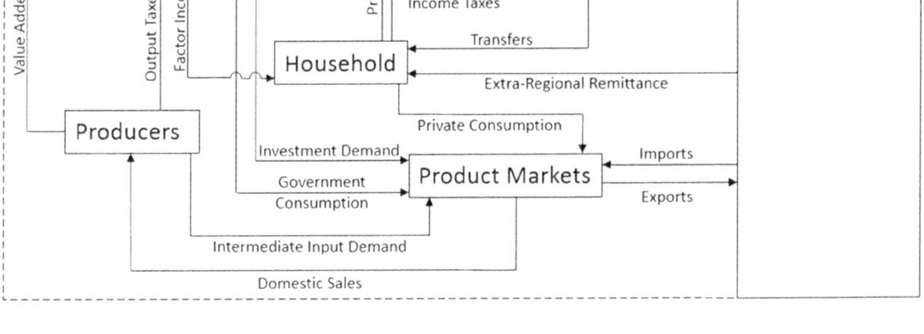

Figure 4-4. Conceptualization of the SAM within CGE.

Rose (2004) identified the basic economic principles of loss estimation, including the distinction between direct and indirect losses, the distinction between stocks and flows, the potential for double counting, and the importance of both property damage and business interruption losses. In addition, Rose (2004) provided a summary of IO and CGE methods in the context of these basic principles.

A notable application of IO modeling to loss estimation is given by Rose and Wei (2013), which estimates the economic impact of a seaport disruption. Rose and Wei (2013) found that output falls as much as $13 billion but that resilience can reduce these impacts by nearly 70%.

Cutler et al. (2016) develop methods for building dynamic, spatial CGE models that can incorporate damage predictions from engineering models. This approach is useful both for quantifying the impacts of disasters across space and time and for evaluating the impacts of investing in the increased resilience of buildings and infrastructure. Fung et al. (2019) apply this approach to quantify the co-benefits of increased resilience for Cedar Rapids, Iowa. They find that a hypothetical shock to productivity results in higher employment and output growth with resilience than without.

Desmet et al. (2018) use a dynamic, spatial CGE model to estimate the economic costs of sea-level rise. They find that permanent flooding due to sea-level rise could reduce global real GDP by an average of 0.19% in present-value terms, with welfare declining by 0.24% due to related migration. Finally, Okuyama and Santos (2014) review the literature on IO and CGE methods for natural disaster loss estimation, highlighting the strengths and weaknesses of each approach.

In general, both CGE modeling and empirical analysis require a large amount of data. A key technical advantage of empirical approaches relative to IO and CGE modeling is the ability to quantify uncertainty in terms of standard errors or confidence intervals. Although CGE models can be data-intensive, they rely largely on aggregate data. Statistical approaches, and causal inference in particular, require data at a finer scale or simply more data in terms of the number of observations and number of predictors (Fung and Helgeson 2017).

4.5 OPTIMIZATION

4.5.1 Portfolio Approach for Decision Making

Deciding the optimal mix of projects or investments requires a portfolio approach. An optimal mix of projects can be selected by their (1) total net benefits or (2) total cost plus expected loss. Each approach represents the dual of the other—the former achieves optimality via maximization, whereas the latter achieves it through minimization. Regardless of the approach employed, they both result in the same solution and result. Often the minimization approach is used because it does not require the estimation of expected benefits; rather, it focuses on costs and avoided losses.

The economic burden model of hazards (EBMH) developed here is adapted from the cost-plus-loss (C + L) model used in wildfire management originated by Sparhawk (1925) and further developed by Donovan and Rideout (2003). Sparhawk illustrated his C + L model graphically depicting the economic tradeoff between investments into primary wildfire protection (e.g., prevention and prefire suppression) and suppression with wildfire losses. Donovan and Rideout reformulated Sparhawk's original C + L model to allow for both primary protection (*presuppression*) and suppression to be independent inputs into the model. (Only primary protection is independent in the Sparhawk model.)

Figure 4-5 shows the Donovan and Rideout approach adapted for resilience. Their presuppression and suppression phases are generalized into resilience preparedness, response, and recovery phases. For illustration purposes, response and recovery expenditures are held constant; thus, as preparedness expenditures increase, disaster-related losses fall. (The figure can be generated holding any two intervention activities constant while allowing the third to vary.) Summing intervention expenditures related to the preparedness, response, and recovery activities with disaster-related losses yield the total societal costs and losses (*economic burden*) associated with natural and human-caused hazards.

The optimal level of resilience spending (i.e., the optimal combination of preparedness, response, and recovery expenditures) is that which

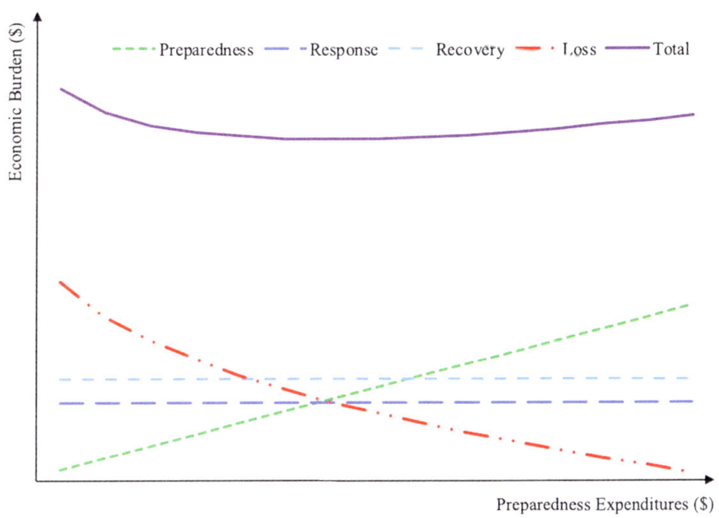

Figure 4-5. Economic burden model of hazards.

results in the minimum economic burden (minimum point based on the total in Figure 4-5). Increases in spending beyond Point A (to the right) are not fully offset by the reduction in damages and are, thus, not economical. This is also true for decreases in spending below Point A (to the left) that result in increases in damages that exceed the marginal savings.

Although the graphical depiction of the EBMH is useful for illustration, it is less so for identifying the optimal resilience spending when preparedness, response, and recovery expenditures are less constrained.

Mathematically, the EBMH model, adapted from Donovan and Rideout, can be written as follows:

$$\text{MIN:}\quad \text{EB} = \sum_{i=1}^{I}\sum_{k=1}^{K}\sum_{t=1}^{T}\left\{\left(w_{i,t}m_{i,t} + f_{k,t}\left(m_{i,t},\ldots,m_{i,t-N_i}\right)\right)e^{-rt}\right\} \quad (4\text{-}15)$$

where

$++w$ = Unit prices for intervention quantity, m, indexed by activity type i (preparedness, response, and recovery) for period t;
$f()$ = Set of k expected loss functions that are a dependent on intervention;
N = Total number of periods in which intervention is effective; and
r = Discount rate.

The first-order conditions to minimize EBMH are follows:

$$\frac{\partial(\text{EB})}{\partial m_{i,t}} = \left\{w_{i,t} + \sum_{k=1}^{K}\sum_{n_i=0}^{N_i}\left(\frac{\partial f_{k,t+n_i}\left(m_{i,t+N_i},\ldots,m_{i,t}\right)}{\partial m_{i,t}}\right)\right\}e^{-rt} = 0 \quad (4\text{-}16)$$

Therefore, optimality conditions imply

$$\sum_{k=1}^{K}\sum_{n_i=0}^{N_i} -\left(\frac{\partial f_{k,t+n_i}\left(m_{i,t+N_i},\ldots,m_{i,t}\right)}{\partial m_{i,t}}\right)e^{-r(t+n_i-1)} = w_{i,t} \text{ for all } i,t \quad (4\text{-}17)$$

Thus, economic burden is minimized when the discounted marginal benefit from reducing losses from hazard events equals the unit price (i.e., marginal cost) of intervention. Again, as was the case with the graphical depiction, at the optimal level of resilience, any increase in intervention will not be fully offset by additional avoided losses, whereas decreases result in damages in excess of the expenditure savings.

Two challenges exist that make identification of the optimal levels of resilience difficult to determine. First, an understanding of the functional relationship between resilience activities and the expected losses is needed. Second, the measurement of all possible resilience activities, their unit prices, and the damages related to hazard events is a significant effort, regardless of the scale (e.g., national, state, and community).

Finally, in reality, the actual set of actionable resilience investment choices may be discrete, meaning that the optimal combination of interventions may not be feasible or within budget. In such cases, the intervention strategy that results in the minimum economic burden, out of the choice set, is the preferred strategy on economic grounds. Note that nonmarket considerations—for example, political, cultural, environmental—or those values that cannot be easily monetized are not part of the economic evaluations and must be considered separately.

Economic evaluation requires the identification and quantification of all associated costs of resilience (e.g., intervention activities) and an understanding of how each intervention activity results in a level (or change) of expected loss. The amount of data can depend on the perspective of the users. For example, a governmental unit may wish to consider all impacts and stakeholders within their governmental jurisdiction, including third parties. An asset owner may only wish to consider their costs and those expected losses that affect them.

It is common to compare proposed options with a baseline for comparison, which may include either a do-nothing or status quo scenario.

4.5.2 Example: Economic Burden Model of Hazards

Table 4-7 presents an example comparison of five resilience decision alternatives. The example highlights the use of the EBMH (minimization) approach and compares the results with a first-cost approach and net

Table 4-7. Example Comparison of Costs and Expected Losses from Decision Alternatives.

Option	Cost	E(loss)	Cost + E(loss)	(Benefit – Cost) | DN	(Benefit – Cost) | SQ
DN	$0	$500	$500	Baseline	n/a
SQ	$100	$200	$300	$200	Baseline
Project A	$125	$150	$275	$225	$25
Project B	$200	$50	$250	$250	$50
Project C	$300	$25	$325	$175	$–25
Project D	$50	$300	$350	$150	$–50

Notes: DN = do-nothing; and SQ = status quo.

benefit (maximization) approach. The five possible alternatives presented are do-nothing, status quo, and three alterative projects (A–C). Each decision incurs different discounted investment costs and discounted expected losses from disasters.

The do-nothing option requires no investment cost into resilience but results in the largest expected losses ($500). The status quo option, which is meant to represent the current level of resilience investment, costs $100 but reduces expected losses by $200 over the do-nothing option. Projects A, B, and C require greater investments than the status quo but result in less expected losses. Project D costs less than the status quo but incurs a larger expected loss.

Comparing the costs plus expected losses, Project B has the smallest economic burden, with the do-nothing option having the largest economic burden. From an economic standpoint, Project B has the best performance. Note some other options have lower costs and some others have lower expected losses, but Project B yields the lowest combination. Compared to the first-cost approach, where the investment with the lowest initial cost is selected (Project D), the EBMH yields a lower total economic burden, as the greater initial investment is more than offset by lower expected losses in the future.

The EBMH (cost-plus-loss) approach is the dual of a net benefits approach, meaning that they should yield the same result. The cost plus loss approach seeks the minimum costs plus losses, whereas the net benefits approach seeks the maximum net benefits (benefits minus costs). In this example, benefits are calculated relative to a baseline. Two different net benefits comparisons are made—one with the do-nothing option as the baseline and a second with the status quo option as the baseline.

Regardless of the baseline selected or evaluation approach used, Project B remains the economically preferred option. The EBMH approach yields the same result as the net benefit approach. An advantage of the EBMH approach over the net benefits approach is that the former requires information regarding costs and expected losses. The net benefits approach requires information regarding costs, expected losses, and in the case of comparing the (non-do-nothing) status quo, it also requires estimates of the do-nothing expected losses. In addition, the do-nothing scenario may be unrealistic or infeasible to construct.

REFERENCES

Ayyub, B. M. 2014. *Risk analysis in engineering and economics*. 2nd ed. Boca Raton, FL: Chapman & Hall/CRC Press.

Ayyub, B.M., and McCuen, R., 2011. *Probability, Statistics and Reliability for Engineers*, Chapman and Hall/CRC Press, Third edition, 2011, Boca Raton, FL.

Brealey, R., and S. Myers. 2000. *Principles of corporate finance*. 6th ed. New York: McGraw-Hill.

Budnick, F. 1998. *Applied mathematics for business, economics, and the social sciences*. New York: McGraw-Hill.

Cavallo, E., S. Galiani, I. Noy, and J. Pantano. 2013. "Catastrophic natural disasters and economic growth." *Rev. Econ. Stat.* 95 (5): 1549–1561.

Cavallo, E. A., and Noy, I. 2009. *The economics of natural disasters: A survey*. IDB Working Paper No. 35. Washington, DC: Inter-American Development Bank.

Copeland, T., and V. Antikarov. 2003. *Real options: A practitioner's guide*. London: Thompson.

Cutler, H., M. Shields, D. Tavani, and S. Zahran. 2016. "Integrating engineering outputs from natural disaster models into a dynamic spatial computable general equilibrium model of Centerville." *Sustainable Resilient Infrastruct.* 1 (3–4): 169–187.

Defusco, R., D. McLeavey, J. Pinto, and D. Runkle. 2001. *Quantitative methods for investment analysis*, 54–56. Baltimore, MD: United Book Press.

Defusco, R., D. McLeavey, J. Pinto, and D. Runkle. 2015. *Quantitative investment analysis*, 44–45. Hoboken, NJ: Wiley.

Desmet, K., R. E. Kopp, S. Kulp, D. K. Nagy, M. Oppenheimer, E. A. Rossi-Hansberg, et al.. 2018. *Evaluating the economic cost of coastal flooding*. Working Paper No. 24918. Cambridge, MA: National Bureau of Economic Research.

Donovan, G. H., and D. B. Rideout. 2003. "A reformulation of the cost plus net value change (C+NVC) model of wildfire economics." *For. Sci.* 49 (2): 318–323.

Fu, C., and J. Gregory. 2019. "Estimation of an equilibrium model with externalities: Post-disaster neighborhood rebuilding." *Econometrica* 87 (2): 387–421.

Fung, J. F., and J. F. Helgeson. 2017. *The resilience dividend: Accounting for co-benefits of resilience planning*. Technical Note No. 1959. Gaithersburg, MD: National Institute of Standards and Technology.

Fung, J. F., J. F. Helgeson, C. O'Fallon, D. Webb, and H. Cutler. 2019. "The resilience dividend in Cedar Rapids." In *Proc., 27th Int. Input–Output Association Conf*, International Input-Output Association, June 30– July 5, Glasgow, Scotland. Accessed March 10, 2021. https://www.iioa.org/conferences/27th/conference.html.

Graham, J. and C. Harvey. 2001. "The theory and practice of corporate finance: Evidence from the Field." *J. Finan. Econ.* 60 (2–3): 187–243.

Okuyama, Y., and J. R. Santos. 2014. "Disaster impact and input–output analysis." *Econ. Syst. Res.* 26 (1): 1–12.

Rose, A. 2004. "Economic principles, issues, and research priorities in hazard loss estimation." In *Modeling spatial and economic impacts of disasters*, edited by Y. Okuyama and S. E. Chang, 13–36. New York: Springer.

Rose, A., and D. Wei. 2013. "Estimating the economic consequences of a port shutdown: The special role of resilience." *Econ. Syst. Res.* 25 (2): 212–232.

Sparhawk, W. N. 1925. "The use of liability ratings in planning forest fire protection." *J. Agric. Res.* 30 (8): 693–792.

Thomas, D. S. 2017. *Investment analysis methods: A practitioner's guide to understanding the basic principles for investment decisions in manufacturing.* NIST Advanced Manufacturing Series 200-5. Gaithersburg, MD: NIST.

Thomas, D. S. 2019. "Monte Carlo Tool." Accessed March 10, 2021. https://www.nist.gov/services-resources/software/monte-carlo-tool.

CHAPTER 5
DESIGNING FOR RESILIENCE

5.1 INTRODUCTION

The concept of resilience, as previously discussed in Chapter 2, is rooted in risk principles. Ultimately, resilience relates to the response of a project or a system in the presence of extrinsic and disruptive events or a series of events, which are usually associated with extreme conditions. Thus, the process of creating resilient systems involves uncertainties that, in many cases, are significant and present major challenges from a design and implementation perspective. For a designer, a resilient system can be understood as one that is engineered to (1) withstand the effects of an anticipated hazard; (2) experience an acceptable level of damage, downtime, or loss of functionality while maintaining a threshold performance level followed by recovery to a target level of performance in a tolerable amount of time; (3) recover as planned; and (4) adapt to new conditions. This target level of postrecovery performance may be smaller than, equal to, or greater than the pre-event performance level. This implies that resilience is not limited to engineering consideration but should also include other considerations related to planning, maintenance, lost services, rehabilitation, life-cycle costs, risk-informed decision making, and, in some cases, socioeconomic impacts. In this process, achieving an appropriate balance among these factors is by no means a straightforward task.

Designing for resilience can only be successful if it is treated as a system's characteristic that can be quantified and measured so that explicit performance targets can be established for the target design life of the system. Standard measures of resilience are, in general, not currently available because they tend to be system and hazard dependent. This latter issue adds a level of complexity that goes beyond the appropriate treatment of uncertainties and their propagation when defining relevant system's characteristics. Moreover, resilience-based engineering and design necessitate clear communication between designers and clients to engineer systems that are adaptable and whose performance exceeds the minimum expectations implicit in design code criteria.

In general, conventional design approaches have involved providing a minimum level of desired reliability, robustness, and redundancy via target strength and serviceability criteria that are required by local and model codes and defined in applicable reference standards. During the last few decades, performance-based design approaches, such as performance-based earthquake engineering, have incorporated strategies for probabilistic assessment of direct economic loss and collapse safety because of earthquakes. In many parts of the country, especially in the US West Coast, the application of performance-based earthquake engineering concepts to life safety and collapse prevention has become more prevalent throughout the years. Significant advancements have been made to the development of standards, and projects have been designed that strive to minimize downtime (i.e., business interruption) and loss of functionality but often without the same level of attention applied to performance targets such as life safety and collapse prevention—the exception might be performance-based design of essential facilities in which downtime and operational performance are critical to the service these facilities provide. Downtime and loss of functionality are fundamental for relating the performance of a single facility to the seismic resilience of a community, region, or country (Almufti et al. 2016). Therefore, an emphasis should be placed on loss of service and recovery as part of the design process to make resilience a feasible and realistic design goal.

This chapter starts with introducing design bases and principles and then offers three case studies that illustrate the application of resilience concepts into design for protecting buildings treated as single projects. The first two case studies pertain to the application of flood-mitigation strategies to protect existing buildings and their content from exposure to flooding caused by coastal storm surges as well as rising sea levels. The third case study deals with the application of resilience-based earthquake design to a proposed high-rise building to enhance its resilience to strong ground shaking. These case studies are by no means comprehensive in terms of covering all components associated with resilience-based design. However, they illustrate, albeit in a rather simple manner, the application of some of the fundamental design bases and principles associated with resilience engineering, as discussed in the next section. The intent is to familiarize the reader with fundamental ways in which resilience engineering has been and can be incorporated into current design practice. These case studies address many, but not all, of the components of resilience engineering (e.g., the treatment of socioeconomic impacts is covered in Chapter 7).

Concepts covered in this chapter are applicable to systems, systems of systems, or networks; however, they require embedding single project designs into the analysis of complex systems covered in Chapter 3. The focus of this chapter is on designing for resilience limited to individual projects (or facilities). The use of the term system in this chapter refers to an individual project.

5.2 DESIGN BASES AND PRINCIPLES

At its most basic level, the engineering of resilient systems relies on two components: understanding and quantifying the most relevant and credible disruptive event(s) (e.g., threats or hazards) and defining the system's (pre- and postevent) performance targets. Defining pre- and postevent performance targets can depend on how critical the function of the system is to the users or the owners of the system. Depending on the system's function, these performance targets can be derived from community-, network-, or structure/site-specific requirements. These targets should account for performance measures and expectations that range from life safety to the loss of services, loss of functionality, acceptable levels of damages, and recovery times. For instance, a performance target specific to buildings may be defining reoccupancy timelines. Another important aspect of the design of resilient systems, which deserves more attention from the engineering community, is the need to account for preparedness and emergency-response plans as part of the process. It is imperative to account for resilience to prevent natural hazards from becoming natural disasters.

Designing systems for resilience involves, at a minimum, compliance with local and model codes. Code requirements are typically established by two means: (1) generally accepted engineering standards that are developed using structural reliability approaches or factors of safety and (2) the collected wisdom and experience of the engineering community related to the performance of infrastructure systems during extreme loading events (e.g., hurricanes, tornadoes, and earthquakes). However, local and model codes only establish *minimum* design requirements and meeting code requirements may not necessarily result in designs that achieve the performance targets established for a resilient system. With this said, in many instances, it has been demonstrated that updates to code requirements have resulted in improved performance of individual buildings when exposed to extreme loading conditions. For example, buildings that were designed in accordance with modern building codes throughout the State of Florida exhibited acceptable performance when exposed to hurricanes (Gurley et al. 2006; Prevatt et al. 2017, 2018). Nevertheless, significant improvements to code requirements and engineering practice on building performance and resilience are still needed. Concerted posthurricane damage investigations yielding observations of good and bad engineering and construction practices were used to inform research studies sponsored by the Florida Building Commission that improved design guide provisions and yielded statistically significant reductions in building damages and loss (Gurley and Masters 2011). Continuing toward increasing the hurricane resilience of Florida's buildings, the Florida Department of Buildings and Professional Regulations (DBPR) has developed a single-place document for Optional Enhanced Construction Provisions for use within the State of Florida. DBPR developed this document following the most recent and devastating Hurricane Michael in 2018 to incorporate information in the document into existing Florida Building Code documents.

As part of the engineering process, it is sometimes challenging to quantify the effect of the hazards, identify the risks associated with system exposure to hazards, especially extreme events, and quantify the consequences of system damage, failure, or loss of functionality. However, identifying and quantifying these factors is necessary to increase infrastructure resilience. For instance, increased levels of resilience in coastal regions can be achieved by incorporating design approaches and mitigation strategies that account for a thorough understanding of both infrastructure vulnerabilities and climate change models, which exhibit significant uncertainties. The earthquake engineering community has made significant advancements in terms of quantifying hazards, improving infrastructure performance, and evaluating the impacts of earthquakes on infrastructure in terms of life safety, repair cost, and downtime. For instance, engineers have the fundamental knowledge to design buildings to regain functionality quickly after a disaster to meet the needs of the community. However, even when modern seismic code provisions are implemented, economic losses from damage and downtime continue to rise. To address this problem, implement performance-based design concepts and help communicate risk to the general public, the earthquake engineering community has recently benefited from the development of systems such as the Resilience-based Earthquake Design Initiative (REDi) rating system and the US Resiliency Council Building rating system (see Armen's comment). For instance, REDi ratings are based on a framework that combines and integrates the roles of structural engineering, risk consulting, and business continuity planning in increasing the confidence that an entity will recover quickly after an earthquake (Almufti et al. 2016). The US Resiliency Council communicates risk via a star rating system that integrates safety, damage, and recovery as relevant metrics. In terms of other hazards and using fire as an example, in the United States, the only permitted alternative to standard fire resistance design for buildings is structural fire engineering (SFE), as constituted in Appendix E of the American Society of Civil Engineers, ASCE 7-16 (ASCE 2017). SFE explicitly evaluates the demand and capacity of structural systems under fire loading in a similar manner as other design loads are treated in structural engineering practice. However, because of persistent misconceptions and lack of industry guidance, designers often erroneously justify structural fire protection variances. In regions exposed to wildfires, it is necessary to develop improved quantification of wildfire hazards that can be translated into estimates of relevant parameters (e.g., radiative flux intensities) that can then be combined with infrastructure vulnerabilities to estimate risk and ultimately increase resilience.

The conclusion here is that although significant progress has been made, resilience-based design should be guided by common principles and a fundamental design basis that facilitates its implementation into practice regardless of hazard and system under consideration. As an example, the Resilient Design Institute (RDI) accessed (2020) recommends the following resilient design principles for the built environment:

- Resilience transcends scales: Buildings, infrastructure, and communities as well as short- and long-term performance objectives.
- Resilient systems provide for basic human needs: Water, sanitation, energy, temperature and humidity, lighting, safe air, health, and food.
- Diverse and redundant systems are inherently more resilient: Diversity improves the ability of a system to respond to interruptions; redundancy provides alternative pathways to mitigate disruptions and facilitate recovery.
- Simple, passive, and flexible systems are more resilient: Simplicity tends to go hand-in-hand with a lower need for maintenance. Flexibility improves adaptation.
- Durability strengthens resilience: Durability involves building practices, building design, and infrastructure.
- Locally available, renewable, and reclaimed resources are more resilient: Resilience increases because our dependence on nonrenewable resources or resources or sources far from the site decreases.
- Resilience anticipates interruptions and a dynamic future: Responding to change provides an opportunity for a wide range of system improvements.
- Find and promote resilience in nature: Natural systems have evolved to achieve resilience and we can learn from them; protection of the natural environment enhances resilience for all.
- Social equity and community culture to resilience: Social aspects of resilience can be as important as physical responses.

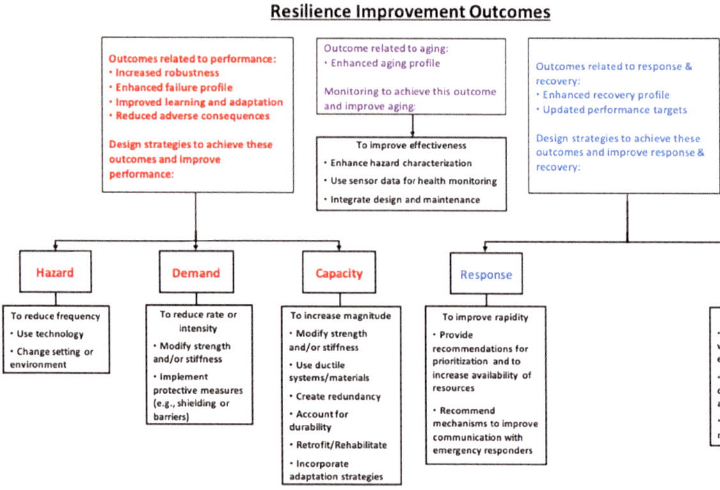

Figure 5-1. Resilience improvement of selected design strategies.

- Resilience is not absolute: Total resilience in the face of all situations is not realistic. Greater resilience can be achieved in steps by implementing what is feasible in the short term.
- Resilience contributes to sustainability: The relationship between resilience and sustainability is discussed in Chapter 1.

It is postulated herein that these design principles can be applied and adapted to specific situations via the design steps identified in Section 5.3.

Figure 5-1 presents various spectra of resilience improvement outcomes that are related to the methodology for resilience analysis presented in Chapter 2 (Figure 2-1) and the information depicted in Figure 2-3. Selected design strategies useful to achieve these outcomes are illustrated, and the importance of monitoring as a fundamental resilience improvement strategy is also highlighted.

5.3 DESIGN STEPS

In general, and consistent with the resilience analysis framework illustrated and discussed in Chapter 2, resilience engineering and design involves the general steps summarized in Figure 5-2 as follows:

- Identify the need for resilience-based design;
- Define target performance goals (IRP 3) (ASCE 2019) because resilience is, in general, discretionary, and defining target performance goals depends on cost–benefit considerations, implying that the process is usually iterative, as shown in Figure 5-2;

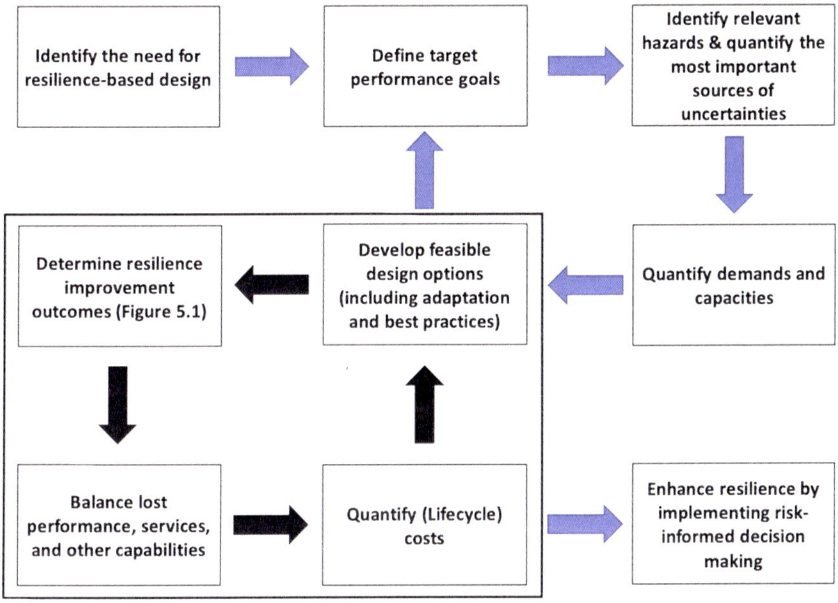

Figure 5-2. Resilience-based design.

- Identify relevant hazards and quantify the most important sources of uncertainties;
- Quantify demands and capacities;
- Iterate over the following four items as presented inside the iteration box in Figure 5-2:
    - Develop feasible design options including adaptation to best practices,
    - Determine resilience improvement outcomes according to Figure 5-1,
    - Balance lost performance, services and other capabilities, and
    - Quantify costs including life-cycle cost; and
- Enhance resilience by implementing risk-informed decision making.

The application of these steps to case studies of individual buildings exposed to flood and earthquake is used for illustration purposes.

5.4 CASE STUDIES

This section illustrates the application of the resilience-based design steps illustrated in Figure 5-2 to the following three case studies:

1. Existing production facility near the US Gulf Coast: Protect critical equipment from damage during a flood so that business operations

may resume shortly after a flood event without the need to repair or replace the equipment.

2. Existing building in Boston: Protect mechanical and electrical equipment critical to the function and safety of the building.
3. New building in San Francisco: Design a high-rise building design using resilience-based design methods (REDi Rating System).

5.4.1 Existing Building Near the US Gulf Coast—Flood Hazard

5.4.1.1 Identify the Need for Resilience-Based Design. A production facility located near the US Gulf Coast houses critical equipment necessary for processing operations and national-level business continuity. The building owner wants to protect critical equipment from flooding during extreme events with the objective of avoiding significant interruption to business operations and production, which would result in sizeable economic losses. Critical equipment that needs to be protected included processing units, electrical switchgear, and a well head providing water when municipal water is unavailable. The basic resilience metric of interest in this example is the duration of service interruption (see Figure 5-3 for general illustrations of critical equipment and building characteristics).

5.4.1.2 Define Target Performance Goals. The specific goal is to mitigate flood-induced damage to relevant equipment and resume operations shortly after a flood event without the need to repair or replace critical equipment. Flood-mitigation concepts for spaces and other areas not containing critical equipment are outside of the scope of the project. Ideally, the definition of performance goals should be consistent with quantifiable resilience metrics. In this case, resuming operations *shortly* after a flood event would need to be translated into an appropriate time interval, which

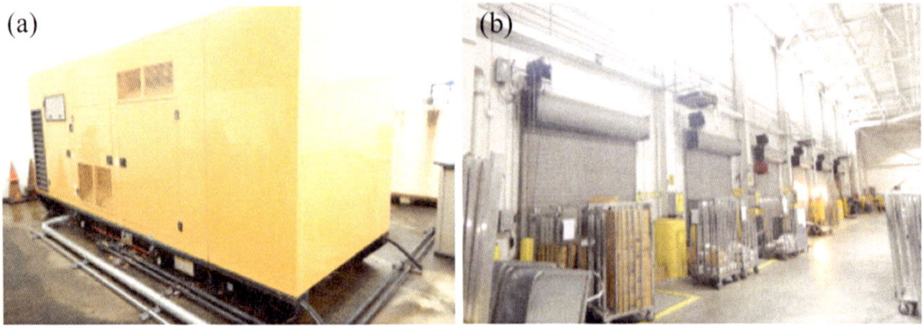

Figure 5-3. Production facility near the US Gulf Coast: (a) critical equipment, and (b) overhead doors.
Source: Simpson Gumpertz & Heger, reprinted with permission.

could range from a matter of hours to days, depending on factors such as the expected relationship between downtime and dollar losses.

5.4.1.3 Identify Relevant Hazards, Quantify the Most Important Sources of Uncertainty, and Quantify Demands and Capacities. Uncertainties are related to hazards, vulnerabilities, building access during and after a flood, and availability of personnel to perform necessary functions because some hazards would affect them, their families, transportation means, and so on. Uncertainties regarding building access are largely because of modeling the building in isolation without consideration of the infrastructure that serves it. Flood hazards primarily because of hurricane-induced storm surges are of interest for this site. Site flood hazards can be evaluated using applicable building codes, standards, and guidelines. Because of location, climate models that incorporate projected flood elevations owing to increased precipitation and sea-level rise (SLR) should be evaluated, including relevant uncertainties, to determine demands. In this case, the likely effects of FEMA's 100 year (base) and 500 year (extreme) floods are considered given the owner's desire to protect critical equipment during extreme events. To quantify the potential effect of storm surges, the National Oceanic and Atmospheric Administration (NOAA) used the Sea, Lake, and Overland Surges from Hurricanes (SLOSH) model to evaluate surges for Category 3, 4, and 5 hurricanes. The SLOSH model is developed by the National Weather Service (NWS) to estimate storm tide and storm surge heights resulting from historical, hypothetical, or predicted hurricanes by considering the atmospheric pressure, size, forward speed, and track data. The parameters can then be used to create a model of the wind field that drives the storm surge. Figure 5-4 depicts the example Category 3 and 5 storm tides and tide level for the high-tide condition in the region of interest.

To assess vulnerabilities, which are associated with existing capacities for an existing building, a site visit, drawing, and additional building information obtained from the building owner and operators are used. This approach gives rise to a qualitative vulnerability assessment using a combination of experience, engineering judgment, and building information. Several vulnerabilities are identified in the structure, for example, crawlspace, loading docks, glass curtain walls, utilities, and insufficient floor/slab structural uplift resistance, that could allow floodwater to enter the building. From this perspective, the vulnerability assessment is limited to identifying the most critical components in the facility and uncertainties inherent in the vulnerability assessment.

5.4.1.4 Develop Feasible Design Options. Given the number of potential entry points for floodwater identified as part of the vulnerability assessment, two different design approaches are considered. The first

(a)

(b)

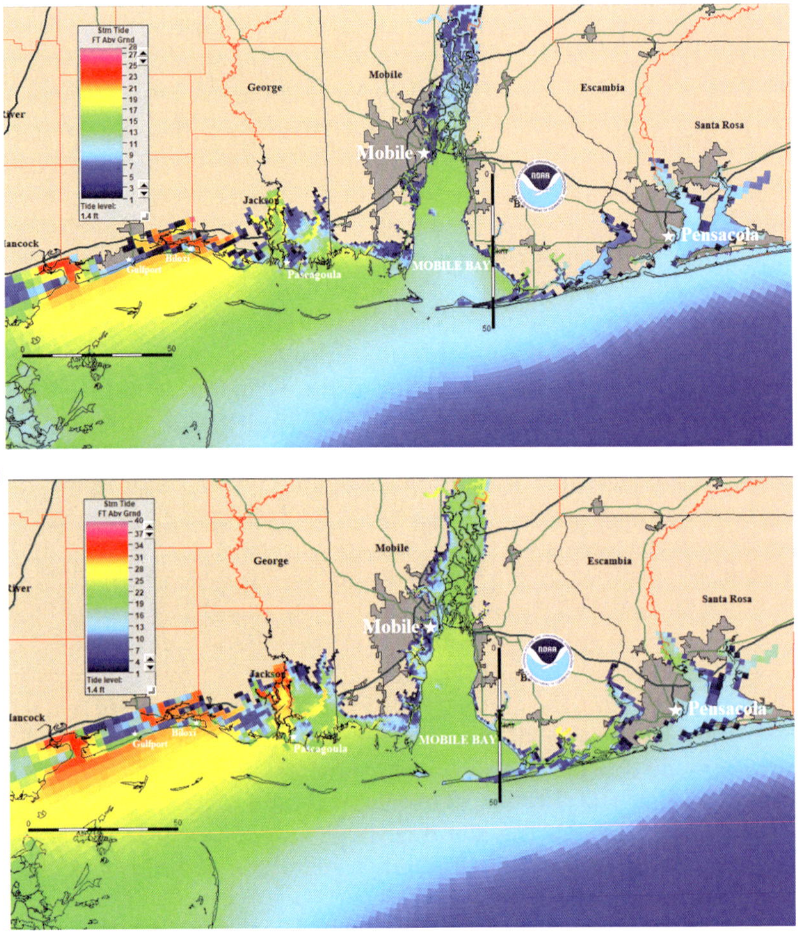

Figure 5-4. Storm tide (NOAA SLOSH model of maximum of maximums): (a) Category 3 storm tide, and (b) Category 5 storm tide.

approach entails creating a flood protection boundary around critical equipment inside the building, which is denoted herein as *local* protection measures. The second approach entails the protection of the perimeter of the production facility, including managing water leakage, which is denoted herein as *global* protection measures. It is anticipated at this point that the successful implementation of either flood-mitigation strategy will have to be supplemented by the installation and maintenance of flood protection measures, the operation and maintenance of emergency equipment, the implementation of a preparedness plan involving emergency personnel and their specific responsibilities during a flood, building access during a flood, evacuation procedures, and training procedures. Two design options are discussed as follows.

5.4.1.4.1 First Design Option: Local Protection Measures. This approach incorporates flood barriers temporarily installed inside the building around critical equipment, elevating key electrical equipment, sealing penetrations and joints, and installing backflow prevention devices on utility lines. At the same time, it is recognized that some water will inevitably leak behind the flood barriers or penetrate the slab within the protected area. Thus, an additional recommendation is to remove this water using sump pumps on emergency power in slab depressions. Leaking rates beneath the barriers can be determined by constructing a mockup of the flood barrier system on an isolated area of floor, allowing more accurate determination of the required pump capacity. An alternative is to size pumps based on a conservatively assumed leakage rate. These local protection measures require relatively minor modifications to the existing building and do not rely on the watertight or structural integrity of the existing walls to resist flooding. One risk associated with this strategy relates to exterior walls not leaking quickly enough to equalize the hydrostatic pressure between the inside and the outside, which could cause the walls to be overloaded structurally. Leaving doors or docks open in selected locations during a flood would allow water to enter the building and avoid unbalanced pressures.

5.4.1.4.2 Second Design Option: Global Protection Measures. This option focuses on protecting the interior and exterior walls that form the perimeter of the production area by using temporary flood barriers at openings (Figure 5-5). Flood water could leak through cracks, joints, and penetrations in the building walls and slabs that are not sealed. Because it is difficult to establish the water tightness or as-built structural capacity of the existing construction until significant flooding occurs, this concept is expected to allow a larger rate of water inflow/leakage into the areas around the critical equipment. Any water that leaks into the building should be removed using sump pumps on emergency power in slab depressions. In this approach, leakage rates through perimeter walls are difficult to estimate because they involve a greater degree of uncertainty when compared with the first design option. Moreover, the level of protection provided by this second design option is limited to the structural capacity of the floor slab/beams and walls to resist hydrostatic pressure and debris impact loads against them, as well as the capacity of the walls to resist forces applied by the temporary flood barriers. Some reinforcement is likely required at the flood barrier connections to the walls, as well as on the walls themselves to resist floods higher than a couple of feet above the finished floor elevation. Once again, this second option involves

Figure 5-5. Example of temporary flood barriers for doors and entrances. Source: FEMA (2013).

a greater degree of uncertainty associated with the effectiveness of the flood protection measures.

5.4.1.5 Determine Resilience Improvement Outcomes. The design strategies identified in the previous section would result in resilience improvements primarily associated with increased robustness, enhanced failure profile, reduced adverse consequences, and enhanced recovery profile (Figure 5-1).

5.4.1.6 Balance Lost Performance, Services, and Other Capabilities. Given the information presented previously, although both local and global measure strategies can be effective in mitigating flood-induced damage to critical equipment, the application of global protection measures is associated with higher uncertainties related to the structural capacity of walls and slabs as well as water tightness of the perimeter of the facility. From a purely target performance point of view, a reasonable tradeoff appears to allow a sizeable amount of water into the building during an extreme flood event as long as the operation of critical equipment is not compromised (second design option).

Additional considerations that would affect performance, service, and capabilities with either design alternative are described as follows:

Sump Pumps: Sump pumps to remove water from protected areas can either be temporary or permanently installed; however, permanent installations tend to be more reliable. In both cases, sump pumps must be connected to an emergency power supply

Table 5-1. Advantages and Disadvantages of Temporary and Permanently Installed Sump Pumps.

Pump Type	Advantages	Disadvantages
Temporary	• Less expensive • Pumps can be moved around as needed during a flood • Some types can collect water of minimal depth	• Requires active human monitoring or water level sensor to turn on/off, which can be unreliable • Requires temporary power connection and discharge piping
Permanently installed	• Permanent switch turn pump on and off as needed; requires no human monitoring • Power connection and discharge piping are already installed • More reliable	• More expensive • Requires sump pit to collect water

to ensure that they will function during a flood event. Advantages and disadvantages of both types are illustrated in Table 5-1.

Temporary Generator: If the building owner decides to bring a temporary generator to the site during or after a flood to provide additional power beyond the capacity of the existing generator to operate production equipment, the following issues should be addressed:

- Locate the generator where the ground elevation is as high as possible.
- Elevate the generator off the ground above the design flood elevation.
- Consider the potential for flooded or impassable roads between the generator origin and the site.
- Connect the generator to the building electrical system via a temporary cable.
- Fuel the generator periodically. If roads are not passable at the time the generator is used, fuel would have to be stored on site.
- Workers may not have access to the building during a flood to install temporary generators, so this risk must be a consideration.

Expected building damage (when a flood elevation is above the finished floor elevation): Flood damage is difficult to predict given that it is highly dependent on flood elevation, duration, water velocity, rate of rise and fall, debris, and wind and wave action. The most likely components to incur damage and require replacement or major repair should be identified during the design process.

5.4.1.7 Quantify Costs. In this case study, cost analysis is based on flood protection concepts for each design option conditioned on a flood intensity greater than the 500 year flood (extreme-event condition). A higher contingency is allocated to the second design alternative because of the greater uncertainty associated with existing construction. The cost for the first design option ranges from $500,000 to $600,000. The cost range for the second design option is $1.1 Million to $1.3 Million. Costs associated with storage or mobilization of portable flood protection equipment or maintenance of permanent flood control equipment, or building repairs following a flood should be considered and may very well bring the aforementioned cost ranges closer to one another. However, these additional cost considerations are not accounted for in this analysis. Although the second option may be more expensive, it may also be more reliable and may limit damage better. Therefore, any decision should not be based on cost alone.

5.4.1.8 Enhance Resilience by Implementing Risk-Informed Decision Making. In this case, the owner is provided with information useful to select a design alternative to limit the loss of service of critical equipment to a minimum after an extreme flood event. From an economic point of view, the owner is provided with two design options that are based on the premise that it is not practical to prevent floodwater from entering the building. Moreover, recommendations associated with flood emergency operations planning, building occupants, and inspection, maintenance, and storage are also provided.

For instance, FEMA P-936 *Floodproofing Non-Residential Buildings* (FEMA 2013) recommends that building owners or operators develop a plan for emergency flood operations. This plan should include installation and maintenance of flood protection measures' installation, operation, and maintenance of emergency equipment such as sump pumps and generators, emergency personnel and their specific responsibilities during a flood, building access during a flood, evacuation procedures, and training procedures.

FEMA also recommends that all building occupants evacuate the building before flood elevations make evacuation impossible. The flood protection measures should not require occupants to remain within the building for the duration of the flood. The deployment time for flood protection measures and the safety of building occupants must be considered in planning for floods.

Inspection and maintenance of flood protection measures are critical to ensure that the measures will provide the expected level of protection during a flood and allow the facility to resume operations shortly after the flood event. Walls, slabs, openings, seals, drainage and pump systems, temporary flood barriers, emergency generators, and all other flood-mitigation components should be inspected on a regular basis and maintained as appropriate so that they are in full working order when an event occurs. Temporary barriers, pumps, hoses, cords, and other components must be stored in a location where they are protected from damage and degradation (sunlight can degrade seals) and accessible on short notice before an event. Electrical and mechanical systems should be tested periodically to ensure that they are in working order.

5.4.2 Mid-Rise Building, Boston—Flood Hazard

5.4.2.1 Identify the Need for Resilience-Based Design and Define Target Performance Goals. The building was constructed in the late 1800s and currently house residences and art studios. Current flood maps demonstrate that the building is vulnerable to coastal flooding and rising sea levels may increase the severity and frequency of flooding. The basement contains various electrical and mechanical equipment that are located several feet below the finished grade elevation and that need to remain operational immediately after an extreme flood event. A mezzanine is positioned between the basement and the first floor.

The building owner previously received a recommendation from an engineering firm to relocate mechanical and electrical equipment to increase the building's flood resilience. Now, the owner is interested in assessing an alternative approach to enhance the resilience of the building to extreme flood events by studying the feasibility of floodproofing the boiler room and other basement areas (in lieu of relocating the equipment).

5.4.2.2 Identify Relevant Hazards, Quantify the Most Important Sources of Uncertainty, and Quantify Demands and Capacities. Most of the uncertainties associated with this study are present in the quantification of the flood hazard at the site, including the potential effect of SLR. To quantify the flood hazard and associated demands, a variety of sources are evaluated, including information on the Flood Insurance Rate Map (FIRM), Flood Insurance Study (FIS), NOAA Tidal Station Records, USACE sea-level change (SLC) curve calculator, the Massachusetts Department of Transportation-Federal Highway Administration (MassDOT-FHWA) Climate Adaptation Pilot Project, the NOAA SLOSH Model, the proposed Massachusetts State Building Code, and the International Building Code.

According to the FIRM, the building is in a Special Flood Hazard Area, which is subject to inundation by the 1% annual chance flood and a base flood elevation of 10 ft North American Vertical Datum of 1988 (NAVD88). Nonstationarity of natural-event processes introduces a level of complexity that is not covered in this example as provided in the ASCE MOP 140 (Ayyub et al. 2018). The FIS summarizes the current flood hazard data collected for a community and provides detailed information about the estimation of the flood elevations. The FIS tabulates the stillwater flood elevations for the bodies of water subject to coastal flooding. Stillwater elevations are the coastal floodwater elevations including storm surge but excluding the effects of wave action. For this building, the FIS stillwater flood elevations are as follows: 1% annual chance (100 year) stillwater elevation: 9.4 ft NAVD88; and 0.2% annual chance (500 year) stillwater elevation: 10.2 ft NAVD88 (1 ft = 30.48 cm).

Historical tidal records from the nearest NOAA Tidal Station are also evaluated, resulting in a mean higher-high water (1 ft = 30.48 cm) of 5.2 ft above mean tide level (MTL) (4.8 ft NAVD88). The highest water level recorded at this station since 1921 was 10.0 ft above MTL (9.6 ft NAVD88). The rest of the top-ten highest recorded water levels were between 8.7 ft and 9.1 ft above MTL (8.3 to 8.7 ft NAVD88). The linear mean sea-level (MSL) trend over 95 years is +0.11 in./year.

Estimated SLC values for Boston are also computed using the USACE SLC curve calculator on the USACE Responses to Climate Change Program website. The USACE SLC curve calculator is the basis for the SLR recommendations in the Boston Redevelopment Authority's Climate Change Preparedness and Resiliency Checklist. For new construction, the Boston Interagency Green Building Committee recommends using at least the intermediate high scenario. Table 5-2 shows the anticipated SLC at future days for NOAA's low, intermediate low, intermediate high, and

Table 5-2. Estimated SLC in Boston Relative to 2016.

Year	Low SLC Scenario (ft)	Intermediate Low SLC Scenario (ft)	Intermediate High SLC Scenario (ft)	High SLC Scenario (ft)
2030	0.12	0.19	0.37	0.56
2050	0.28	0.53	1.08	1.71
2070	0.45	0.94	2.02	3.26
2100	0.70	1.69	3.87	6.36

Note: (1 ft = 30.48 cm).

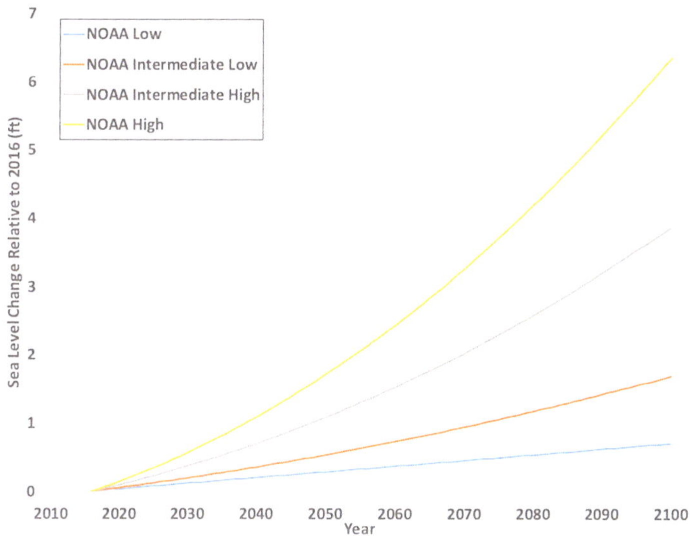

Figure 5-6. Estimated sea-level change at NOAA Station of interest (Boston) relative to 2016 sea levels.

high SLC projection models. The low SLC scenario is only based on the 2016 measured sea-level trends, whereas the other scenarios include the effects of phenomena such as increased ocean warming and ice sheet loss. The anticipated SLC values only incorporate increases in MSL and do not include increased wave setup and wave heights because of SLC. Figure 5-6 shows the estimated SLC curves for Boston from 2016 to 2100.

The Massachusetts Department of Transportation (MassDOT)-FHWA completed a study including SLC, storm surge, and wave modeling for a region encompassing most of the coastal areas in metropolitan Boston. Based on this study, Table 5-3 shows the approximate flood hazard for the building site under consideration obtained from the overall risk maps included in the MassDOT-FHWA Pilot Project Report.

As demonstrated in the MassDOT-FHWA Pilot Project Report, the coastal flood elevations and inundation probabilities will likely increase in the future because of SLC. Although the probability of inundation at the building is approximately 1% chance of exceedance per year, the MassDOT-FHWA model shows that the annual probability could increase to approximately 2% by 2030 and reach as high as 50% by 2070 (assuming high SLC) or 2100 (assuming intermediate high SLC). By approximating the trend of annual inundation probability between these data points, the annual and cumulative chance of inundation curves with SLC can be developed. The annual chance of inundation in the year 2036 is approximately 4%, assuming the high SLC scenario. The resulting

Table 5-3. Estimated Flood Hazard Data from MassDOT-FHWA Pilot Project Report.

Year and SLC Scenario	Approx. Annual Chance of Inundation from Coastal Flooding (%)	Approx. Flood Depth with 1% Annual Exceedance Probability
Present day	1	1 ft (10 ft NAVD88)
2030, High SLC scenario	5	2 ft (11 ft NAVD88)
2070, High SLC scenario/ 2100, intermediate high SLC scenario	50	5 ft (14 ft NAVD88)

Note: (1 ft = 30.48 cm).

cumulative likelihood of inundation until 2036 is approximately 30%. Note that these cumulative probabilities are computed from estimated SLC probabilities that contain a high degree of uncertainty.

Finally, storm surge data from the NOAA SLOSH model that computes storm surge elevations from simulated hurricanes are also evaluated. The SLOSH model is the basis of the Storm Surge Inundation Maps used by USACE for hurricane evacuation planning. SLOSH reports a *maximum of maximums* storm surge elevation that envelopes many different simulated hurricanes with many possible storm tracks. However, NOAA does not assign probabilities to the SLOSH model results, which are generated by combining several storms with different probabilities of occurrence. The SLOSH model uses an average ground surface elevation to compute the floodwater depth for a given area. The elevations obtained from this model can be considered worst-case, high water values based on the storm category (Table 5-4).

On the vulnerability side, which relates to capacities for this existing building, the assessment is based on field observations of the basement mechanical and electrical rooms and the building exterior (Figure 5-7). Information is also collected from the building superintendent.

5.4.2.3 Develop Feasible Design Options. Proposed mechanical upgrades associated with relocating the existing equipment do not represent a substantial improvement to the building. Therefore, the governing building codes do not require that the building be brought into compliance with current provisions for flood protection. In this case, potential flood protection strategies implemented at the building would be up to the discretion of the owner.

Table 5-4. Estimated NOAA SLOSH Model Storm Surge Elevations.

	Maximum of Maximums Storm Surge Elevation (ft NAVD88)		
Hurricane Strength	If Hurricane Occurs at Mean Tide Level	If Hurricane Occurs at 2 ft above Mean Tide Level	If Hurricane Occurs at 5 ft above Mean Tide Level (High Tide)
Category 1	6.3	8.3	11.2
Category 2	11.9	13.1	15.9
Category 3	15.6	17.2	20.1
Category 4	20.2	21.5	23.3

Note: (1 ft = 30.48 cm).

Figure 5-7. (a) Boilers and expansion tank in the boiler room, (b) main switchboard, wall-mounted meter stacks, and wall-mounted panels in the electrical room, (c) fire pump and fire pump controller in the fire pump room, and (d) exposed south foundation wall in the basement.
Source: Simpson Gumpertz & Heger, reprinted with permission.

The selection of a depth of flood elevation (DFE) from the available sources should be based on the level of flood risk that the owner is willing to accept. The building code and FEMA flood elevations are based on a 1% annual chance of storm; however, increased resilience typically implies a design to a less frequent event, for example, the 0.2% annual chance storm. Based on the available sources of flood hazard data evaluated, the following elevations are potential DFEs that could be used in the design of flood-mitigation measures at this site (1 ft = 30.48 cm):

- DFE of 10 ft NAVD88 would provide present-day protection for a 1% annual chance flood. This can be interpreted as the minimum level of flood protection. The risk of flooding will likely increase in the future because of SLC.
- DFE of 11 ft NAVD88 would provide, based on data and knowledge available at the time, protection for an approximately 0.2% annual chance flood in the present day and a 1% annual flood in the year 2030.
- DFE of 14 ft NAVD88 would provide, based on data and knowledge available at the time, protection for an approximately 1% annual chance flood in the year 2070 (assuming high SLC) or 2100 (assuming intermediate high SLC). This represents a relatively long-term level of flood protection.

Alternative DFE could be selected to provide intermediate or higher levels of flood protection.

Table 5-5 depicts the approximate floodwater depths within the basement of the building for the possible DFEs.

Given that the minimum grade elevation at the face of the building is 8.4 ft NAVD88, the basement floor elevation is 4.8 ft NAVD88, the bottom elevation of the deepest mezzanine level framing members is 13.8 ft NAVD88, and the top of the mezzanine floor is at 16.5 ft NAVD88, only the

Table 5-5. Approximate Floodwater Depths.

Depth of Flood Elevation	Maximum Floodwater Depth At Exterior Face of Building (ft)	Maximum Floodwater Depth in Basement (ft)
10 ft NAVD88	1.6	5.2
11 ft NAVD88	2.6	6.2
14 ft NAVD88	5.6	9.2

Note: (1 ft = 30.48 cm).

14 ft NAVD88 DFE has the potential to submerge the bottom of the deepest mezzanine level framing members (1 ft = 30.48 cm). None of the DFEs in Table 5-5 exceeds the elevation at the top of the mezzanine floor.

In this case, three basic alternatives are available to the owner:

- First option: Relocate the equipment (at higher elevations)—original approach recommended before this study was conducted.
- Second option: Dry floodproof only a portion of the basement (with consideration given to wet floodproofing the areas of the basement that are not floodproof).
- Third option: Dry floodproof the entire basement.

5.4.2.4 Determine Resilience Improvement Outcomes (Figure 5-1). The design options described in the previous section would result in resilience improvements identified as follows.

- First option: Increased robustness, enhanced failure profile by increasing ductility, for example, reduced adverse consequences, and enhanced recovery profile.
- Second option: Increased robustness and reduced adverse consequences.
- Third option: Increased robustness, enhanced failure profile, reduced adverse consequences, and enhanced recovery profile.

5.4.2.5 Balance Lost Performance, Services, and Other Capabilities. Dry floodproofing will require substantial structural modifications to the existing basement floor system to provide resistance to the hydrostatic forces (including buoyancy of the slab). The magnitude of the hydrostatic forces and the extent of structural modifications will be strongly dependent on the selected DFE. If the owner is to pursue dry floodproofing in a portion of the basement, she/he should consider selecting the highest possible DFE (i.e., near the ceiling of the basement) to reduce the present-day and future flood risk as much as possible.

Dry floodproofing the entire existing basement structure because of the age of the building and the number of existing leakage sources (e.g., cracks, joints, penetrations, windows, doors, and unsealed electrical conduits) is not a practical alternative. Because dry floodproofing the entire basement is not required anyway by the building code, it might be more prudent in this case to allow the floodwater to enter the basement and protect some or all portions of the basement using selective floodproofing methods. Under this scenario, consideration should be given to wet floodproofing all areas of the basement that are not floodproofed. Potential wet floodproofing measures include the following:

- Installation of pressure relief valves in the existing slab-on-grade or in the exterior foundation walls within 1 ft of the top of the slab to allow water to enter the basement and lower the buoyance pressure on the slab (1 ft = 30.48 cm).

- Replacement of the existing gypsum board partition walls, floor coverings, and other architectural components with flood damage-resistant building materials.
- Replacement of the wall cavity insulation with flood damage-resistant insulation.

5.4.2.6 Quantify Costs. Dry floodproofing the structure could be achieved via considering the scenarios in the following. In all scenarios, the concrete components must be designed and constructed to be substantially impermeable:

- Low relative cost: New structural slab with interior flood walls—This option relies on the existing structure to resist the gravity loads and buoyancy loads.
- Medium relative cost: New isolated, thickened slab-on-grade with interior flood walls—This alternative relies on thickened slab and walls to resist the buoyancy loads and relies on soil below to resist gravity load.
- High relative cost: New slab supported on micropiles with interior flood walls—This alternative could be selected if none of the aforementioned ones can be implemented.

Temporary or permanent premanufactured flood barriers could also potentially be used instead of concrete walls. However, temporary flood barriers require human intervention and hence have greater risk than permanent walls or barriers.

It should be noted that a life-cycle cost analysis will give a different and better basis for decision making than initial cost.

5.4.2.7 Enhance Resilience by Implementing Risk-Informed Decision Making. The scope of any proposed equipment replacement after the basement is inundated following a flood event does not constitute a substantial improvement to the building, and the flood provisions in the building code are not applicable in this case. However, to increase flood resilience, a flood-mitigation strategy should be implemented given that the basement mechanical and electrical equipment is at risk of flood damage and this risk is expected to increase significantly with time.

An appropriate DFE should be selected in addition to the extent of flood protection based on the level of flood risk and downtime that the owner is willing to accept. It is evident that the proposed plan to elevate the boiler room equipment to the roof is the most reliable and practical method of flood protection—it provides the lowest present-day and future risk of inundation.

As discussed, dry floodproofing all or a portion of the existing boiler room is a feasible solution, but it requires significant structural modifications. Moreover, dry floodproofing the electrical room and the emergency power room would require significant electrical work and downtime.

In general, it is recommended that any critical basement level equipment or building systems that are to be replaced or augmented in the future should be installed above the predicted 100 year flood elevation, including the effects of SLC.

5.4.3 High-Rise Building, San Francisco—Seismic Hazard

This case study is adapted from Almufti et al. (2016).

5.4.3.1 Identify the Need for Resilience-Based Design. A 56-story mixed-used (office and residential) high rise (Figure 5-8), which exceeds the height limitations of the building code, is located in San Francisco and designed incorporating performance-based engineering, as required by the San Francisco Building Code Administrative Bulletin 083, in combination with the REDi system. In San Francisco, this type of building should be designed based on performance-based design concepts with minimum

Figure 5-8. High-rise building during construction.
Source: Resilient Design Institute (2020).

code performance objectives related to a low probability of collapse because of maximum credible earthquake (MCE) level ground motions, defined as the largest earthquakes that can reasonably be expected. After engineers discussed with the owner the expected seismic-induced dollar losses and downtime associated with high-rise design based on code-based and performance-based design approaches, the building owner decided to adopt resilience-based design concepts via the implementation of the REDi system. The primary seismic resilience metrics are associated with immediate reoccupancy, which is particularly important for residents, and functional recovery once utilities are restored.

5.4.3.2 Define Target Performance Goals. In this case study, resilience-based performance levels and objectives are associated with median repair costs that are less than 5% of the total building value, a median reoccupancy time of one day (controlled by the postearthquake inspection), and a functional recovery time (i.e., an acceptable timeframe for tenants) of less than a month assuming that utilities are restored.

5.4.3.3 Identify Relevant Hazards and Quantify the Most Important Sources of Uncertainty/Quantify Demands and Capacities. As stated previously, the hazard of interest is seismic hazard. Major sources of uncertainty are related to the seismic hazard at the site as well as analysis models and methods. On the hazard side, the implementation of the REDi in this case, that is, achieving the target performance levels and objectives for this building, involves performance criteria in which nonstructural and structural elements should remain essentially elastic (i.e., demands should not exceed the limit of elastic behavior) and operational for a 475 year earthquake. This implies the need to develop ground motions at the 475 year mean-return period to conduct nonlinear response history analysis and verify that mean nonstructural- and structural-component demands remain in the elastic range. The selection of appropriate analysis models and methods, especially for nonlinear analysis, may involve significant uncertainty.

5.4.3.4 Develop Feasible Design Options. Seismic design features that can be incorporated in the design process for structural elements to achieve the target performance goals include the addition of viscous dampers into the external megabrace system to reduce demands and using uplifting megacolumns at the base of the structure to reduce demands in the foundation and in the columns themselves. Some localized and limited yielding can be accepted in structural components when ductile materials are used, and the mean demands are elastic.

For nonstructural components, design options included a facade system that should remain water-right at lateral drift ratios that exceed the mean drift ratios obtained from the set of 475 year ground motions; one elevator that services the whole building and that satisfies the California Building Code standards for elevators in hospitals; stairs designed to maintain dead and live loads with minimal damage during MCE-level ground motions; and anchorage of architectural, mechanical, and electrical components to remain in the elastic range, on average, during the 475 year ground motions.

In addition, a risk assessment, which includes an evaluation of financial loss and downtime, should be conducted to verify that the performance goals are achieved.

5.4.3.5 Evaluate Resilience Improvement Outcomes (Figure 5-1). The resilience improvement outcomes associated with the design strategies described in the previous section for structural and nonstructural components relate to increased robustness, reduced adverse consequences, and enhanced recovery profile.

5.4.3.6 Balance Lost Performance, Services, and Other Capabilities. Given the performance goals and design strategies presented previously, it is evident that from a functional recovery point of view, resilience-based design options provide clear advantages over the performance criteria based on collapse prevention under the maximum considered earthquake. These advantages concentrate primarily on achieving enhanced resilience objectives related to the mitigation of dollar losses and downtime. An important component that still needs to be addressed is the quantification of costs, which is the subject of the next section.

5.4.3.7 Quantify Costs. Experience and research (Almufti et al. 2016) have shown that the cost premium associated with enhancing the performance of a tall building designed to meet the minimum code objectives to a resilience-based seismic design approach varies between 0% and 5%. The building owner should also consider the premium on design fees related to resilience-based design, which in many cases may be controlled by the need to perform nonlinear response history analyses. This latter premium on design fees becomes a moot issue when the alternative to resilience-based design is performance-based design because both methods involve nonlinear response history analyses. Additional design fees can be related to risk assessment and the peer-review process. Some of the aforementioned costs can be offset, for example, if the building owner pursues reduced earthquake insurance costs or if the developer incorporates seismic resilience components of the project into his/her marketing strategies.

5.4.3.8 Enhance Resilience by Implementing Risk-Informed Decision Making. In addition to the information presented in the previous section, for this building, enhancing and managing resilience would benefit from the implementation of contingency planning and preparedness approaches by the building owner. This implementation would help ensure that reoccupancy and functional-recovery goals are met.

5.5 SUMMARY

This chapter covers proposed design bases, principles, and steps related to the implementation of resilience-based design strategies using individual buildings located in the US Gulf Coast, East Coast, and West Coast of the United States as case studies. The relationship between design strategies and resilience improvement outcomes, which are part of the resilience analysis framework included in Chapter 2, is illustrated and briefly discussed in each case study. The information presented in this chapter is an attempt to highlight important concepts useful to design for resilience with a focus on two hazards only: flood and earthquake. However, the principles and steps addressed in the case studies can be readily transferable to other hazards.

REFERENCES

Almufti, I. A., A. Krebs, E. Long, N. Youssef, O. Hata, and G. Deierlein. 2016. "Practical insights from project implementation of the REDi rating system." In *Structural Engineers Association of California Conf*, Maui, Hawaii. Accessed March 10, 2021, https://www.seaoc.org/store/viewproduct.aspx?id=8735232

ASCE. 2017. *Minimum design loads and associated criteria for buildings and other structures*. ASCE 7-16. Reston, VA: ASCE.

ASCE. 2019. *Resilience-based performance: Next generation guidelines for buildings and lifeline standards*. IRP 3. Reston, VA: ASCE.

Ayyub, B. M., M. Medina, T. Vinson, D. Walker, R. N. Wright, A. AghaKouchak, et al. 2018. "Climate-resilient infrastructure: Adaptive design and risk management." In *Committee on adaptation to a changing climate*, MOP 140, edited by B. M. Ayyub. Reston, VA: ASCE. Accessed March 10, 2021. https://ascelibrary.org/doi/book/10.1061/9780784415191

FEMA. 2013. *Floodproofing non-residential buildings*. FEMA P-936. Washington, DC: FEMA.

Gurley, K., R. H. Davis Jr., S.-P. Ferrera, J. Burton, F. Masters, T. Reinhold, et al. 2006. "Post 2004 Hurricane field survey—An evaluation of the relative performance of the standard building code and the Florida building code." In *ASCE Structures Congress 2006*, edited by B. Cross and J. Finke. Reston, VA: ASCE.

Gurley, K. R., and F. J. Masters. 2011. "Post-2004 hurricane field survey of residential building performance." *Nat. Hazard. Rev.* 12 (4): 177–183.

Prevatt, D. O., K. R. Gurley, D. B. Roueche, and G. Wong-Parodi. 2017. *Survey and investigation of buildings damaged by Category III Hurricanes in FY 2016–17—Hurricane Matthew 2016*. Final Rep. Project# 0004029. Gainesville, FL: Florida Dept. of Business and Professional Regulation (Building Codes and Standards), University of Florida.

Prevatt, D. O., K. R. Gurley, D. B. Roueche, and G. Wong-Parodi. 2018. *Survey and investigation of buildings damaged by Category II, III, IV and V Hurricanes in FY 2017–18—Hurricane Irma 2017*. Final Rep. Gainesville, FL: University of Florida.

RDI (Resilient Design Institute). 2020. "181 Fremont San Francisco." Accessed March 10, 2021. https://www.skyscrapercenter.com/building/181-fremont/664.

CHAPTER 6
COMMUNITY SOCIOECONOMICS

6.1 MOTIVATING FACTORS AND BENEFITS

Infrastructure are the physical, technical, and organizational structures, facilities, and technologies (e.g., buildings, roads, power supplies, broadband, and so on) needed to maintain our way of life. Looking at the context definition in Figure 2-1, enabling an accurate analysis of community resilience requires a depiction of the physical infrastructure systems as well as the population, its sociocultural norms, and community-set goals. When infrastructure is disrupted, it has a systemic and unequal impact on society. Therefore, any infrastructure investment should aim to prevent such disruption by being adaptable to future change, including climate change, economic growth, population and demographics, or globalization. As a guiding framework, infrastructure should be viewed as a supporting societal function. As such, the goal of this chapter is to articulate that relationship and provide social and economic metrics that can be used to quantify an infrastructure project's impact on the surrounding community. In addition, infrastructure providers need to be aware of strategic risks and potential opportunities coming from technological stresses such as automation, artificial intelligence, and cybersecurity.

Disasters magnify social inequalities induced by the placement and access to safe and diverse infrastructure in communities (Chang 2016, Hendricks 2017). When a community's social and economic institutions are not directly and holistically considered for the whole community in infrastructure investment and design, the most vulnerable get overlooked (Hendricks et al. 2018, Hamideh and Rongerude 2018,). Looking at the US, social inequality and disparities are prevalent across the country (Fothergill and Peek 2004, Shirley et al. 2012). After each disaster, the areas hit the hardest are those in which unemployment, poverty, low income, and lower levels of education were already factors (Sutley 2018).

This is not only a problem for the US. Some have argued that the world is entering into what has been termed a VUCA (volatile, uncertain, complex, and ambiguous) future. This VUCA state is defined by a combination of stress factors such as climate change, increasing occurrence and intensity of extreme events, resource scarcity, increasing wealth inequality, an aging population, increasing disease, housing shortages, population and traffic congestion, poor air and water quality, the replacement of labor forces through automation, and dysfunctional political governance. Together, this instability operates to challenge the fixed notions of performance, operations and maintenance, and reliability that have historically defined civil engineering projects (Chapin et al. 2010).

The relationship between shocks and stresses is not straightforward. Many of these stress factors increase the impact and frequency of shock incidents or make the systems that support our way of life more susceptible to them. Climate change is undoubtedly gradual, but a changing climate is associated with an increased incidence and severity of a variety of shocks. For instance, high summer temperatures can tip into periods of acute and lethal heat stress, which can be exacerbated by energy and transport disruptions and can lead to a variety of social costs (Hondula et al. 2015). Gradual change can create new thresholds for sudden shocks to occur—with cascading consequences. When shocks do occur, the interdependent infrastructure systems, such as water and wastewater, electric power, transportation, and telecommunications systems, that enable everyday life may fail or be compromised. These events disrupt the continuity of services that most people take for granted (Kruchten et al. 2007). As observed after 2017 Hurricane Irma, service outages and restoration times do not occur equally across a community, leading to differential impacts, suffering, and recovery experiences (Chakalian et al. 2018, Mitsova et al. 2018). Chapter 4 highlighted resilience socioeconomics and noted that there is a significant community cost associated with losing social capital (Aldrich and Meyer 2015). Social capital is related to social vulnerability but is uniquely defined as the networks of relationships individuals have that provide them access to information, resources, and decision-making power. Thus, when social capital is lost in a community, so too is community cohesion and social equity.

Severe shocks have consistently led to unequal impacts, associated costs, and differential recovery rates in cross sections of communities—often hitting the most disadvantaged and vulnerable hardest (Peacock et al. 2014, Sutley 2018). An example of such an impact is the City of Lumberton, North Carolina, which is a triracial community with median income far below the US average and nearly three times the national poverty rate. Lumberton suffered extensive flooding following Hurricane

Matthew in 2016 and flooding again following Hurricane Florence in 2018. The flooding originated when water from the Lumber River backed up and came around the levee system, which had previously been identified as deficient and subject to future flooding. In flood-prone areas, it is quite common to use levees to rationalize building in a floodplain, where it is low- to marginally middle-class households who reside in these areas, even though decades of disaster scholars have pointed out numerous failure of levees systems. Thus, as would be expected, 13 years after the levee inspection in Lumberton, North Carolina, the city was catastrophically flooded, a preventable disaster caused by deficient infrastructure. Inundation maps of Lumberton, North Carolina, overlaid with census data to identify vulnerability hotspots demonstrated that the flood did not impact the city or its residents equally. Rather, those residents in low-income neighborhoods and primarily racial minority and ethnic neighborhoods experienced the highest inundation levels. A recent study of Lumberton, North Carolina, after these disasters illustrated what the social sciences have shown for years that population stability (i.e., dislocation of households) is not only a function of flood inundation and damage but also a function of race, ethnicity, and housing ownership status (renter versus owner) (van de Lindt et al. 2018, 2020). The analysis showed that minority race and ethnic households were as much as three times more likely to dislocate than non-Hispanic White households, even when the same level of building damage was present. This provides one of many examples demonstrating that it is more apparent than ever, community resilience is a multidisciplinary endeavor which cannot be solved by engineers alone.

Nevertheless, infrastructure projects have historically been planned and executed based primarily on economic considerations (see discussion in Chapter 4). It is becoming increasingly evident that monetizing should only play a part of the role in the project decision-making process and a multifaceted, or holistic, approach must be taken into account that supports community-set goals for accommodating the needs of society. A review of literature that focuses on the relationship between physical systems, social science, and economics can be found in Koliou et al. (2020), and some examples of the effect of discounting social losses, often referred to as indirect losses, can be found in Sutley et al. (2016a, b, 2017). Finally, for the interested reader, perhaps the largest systematic effort on monetizing disaster losses in the US can be found in Multihazard Mitigation Council (2018), which measured a national benefit of $11 in postdisaster savings for every US dollar invested in predisaster mitigation that included estimates on economic losses generated from the workforce having post-traumatic stress disorder. That study attempted to account for not only direct saving but also indirect saving attributed to socioeconomic factors not considered in the earlier effort.

Historically, risk has been viewed as economic from a project standpoint, but a fully risk-informed decision support platform does not attempt to quantify the supporting information as a single unit, or single monetary value; rather, risk information should be presented as a dashboard providing multiple metrics. This enables the stakeholders to actually make a risk-informed decision in terms of possible disruptions to the physical infrastructure, as well as the social and economic institutions the physical infrastructure supports.

6.2 SOCIOECONOMIC NEEDS AND METRICS

Physical infrastructure systems are part of a broader system that includes people, governance, and economy (Cimellaro et al. 2016). Indeed, there are seven capitals within a functioning community, namely, human, cultural, social, political, financial, natural, and built (Ritchie and Gill 2011). In this framework, physical infrastructure represents only one asset, or capacity, in a community and should be designed from the perspective of supporting the other six capitals. Physical infrastructure systems contribute to delivering common community values such as security, economic growth, quality of life, and sustainability. Resilience creates, protects, and enables value to be continuously delivered within this interconnected system.

It is important that infrastructure should not be designed in isolation of the community it resides and for designers to understand the social and economic purpose and outcomes that infrastructure and buildings support so that this purpose can be maximized as part of the design process. It is also important that this consideration can be done consistently and repeatedly across project teams, infrastructure types, and communities (Johansen et al. 2017); thus, metrics are proposed here to aid in standardizing the incorporation of social and economic systems in infrastructure projects. Some examples of social, economic, and physical metrics are listed in the following accompanied by tabulated suggestions for design considerations to capture social and economic metrics throughout the various stages of infrastructure projects.

- Social metrics account for the quality of life factors for the culture within which infrastructure resides and provide an outlet of expression for the arts and humanities inherent to that culture. They could include metrics in the following categories:
  - Health and well-being,
  - Community cohesion and inclusivity,
  - Social equity,
  - Diversity,

  - Mobility and accessibility, and
  - Safety.
- Economic metrics account for the contribution that infrastructure spending has on the financial well-being of the community as a whole and on individuals, as well as specific jurisdictions and enterprises. Growth of both individual wealth and community wealth is the key objective that such metrics measure. They could include metrics in the following categories:
  - Productivity,
  - Employment,
  - Economic diversity,
  - Economic growth, and
  - Home ownership.
- Physical metrics account for the traditional engineering metrics used to evaluate public and private infrastructure and are intended to ensure that materials and projects are used in such a way as to prevent damage or danger to the community and its social, governance, and economic fabric. They could include metrics in the following categories:
  - Performance,
  - Functionality,
  - Adaptability,
  - Recovery time, and
  - Maintenance.

In line with the first three steps from the six-step process (Figure 6-1) from the community resilience planning guide (NIST 2015) and decades of academic literature (Meyer et al. 2018, Levine et al. 2007, Burby et al. 2000), communities should form collaborative teams with representation across all sectors and capitals and together set long-term community goals, infrastructure performance goals, and objectives to reach these goals. Community-level collaboration should not stop there but should continue to execute all six steps shown in Figure 6-1, including developing a plan, executing the plan, and evaluating and modifying as the community, its hazards, and socioeconomic structure may evolve over time.

In line with this, new infrastructure projects should integrate the community goals accommodating representatives across the entire community. Community socioeconomic goals can be enhanced at the project level through consideration at the appropriate stage (e.g., preplanning, planning, design, construction, and operation). To aid in this process adoption, Table 6-1 aligns the social and economic metric categories with example community goals and outcomes, demonstrating how these can be executed and incorporated into infrastructure projects.

Figure 6-1. National Institute of Standards and Technology six-step process to planning for community resilience.
Source: NIST (2015).

Metrics used within a resilience assessment (Field et al. 2017) for a community or project are quantitative in nature, but qualitative considerations can enter into the decision-making process by providing additional information on which to judge and weigh a dashboard of metrics. Referring to the overarching framework in Figure 2-1, this dashboard would be critical for the last step, risk-informed decision making for resilience management, by integrating what is learned from the physical, social, economic, sustainability, and governance considerations. Each metric may have a different level of importance and different

Table 6-1. Meeting Community Goals through Infrastructure Projects.

Preplanning	Planning	Design	Construction	Operation	Health and Well-being
Outcome: Green community spaces					
	X				Provide appropriate green space and consider untapped opportunities for public realm space
				X	Engage other local groups and charities to take an active role in long-term maintenance of public and open spaces
		X			Provide facilities to encourage the use of sustainable transport options such as bicycles, electric vehicles, and walking to reduce vehicle emissions in community and, in some cases, supporting health and fostering community cohesion
		X	X		Avoid using harmful building materials
Outcome: Healthy air quality					
		X			Provide facilities to encourage the use of sustainable transport options such as bicycles, electric vehicles, and walking to reduce vehicle emissions in community
		X	X		Avoid using harmful building materials
Outcome: Good physical health					
	X				Incorporate leisure and gym facilities into the development
		X			Provide thermal comfort, good acoustics, and natural daylight and ventilation for the development
				X	Set up carpooling groups and bicycle hire schemes
Outcome: Rapid recovery postevent					
				X	Develop and exercise robust response and recovery plans, including communicating hazard risks and identifying vulnerability hotspots

Preplanning	Planning	Design	Construction	Operation	Community Cohesion and Inclusivity
Outcome: Strong local ownership of the project/development/local risks					
X					Communicate local risks and what your development is being designed to achieve
		X			Consider opportunities for community-led development
	X	X	X		Encourage and facilitate local community engagement throughout project
X	X				Align development with community values
				X	Gather community feedback on the development
Outcome: Thriving social networks					
	X	X			Provide high-quality community facilities as part of the development, that is, garden, meeting space, leisure facilities
		X			Design area to encourage social interaction between occupiers and local community
Outcome: Strong sense of community culture and heritage					
	X				Avoid demolition of assets with community value and protect and enhance historic buildings
		X			Design with sensitivity to local historical context
		X			Consider design features that bring character and local pride to the development

Preplanning	Planning	Design	Construction	Operation	Social Equity
Outcome: Fair and just project teams					
X	X	X	X		Provide opportunity for input and questions from all project members
X	X	X	X		Ensure diversity in experience, gender, background, and professional expertise on project team

(*Continued*)

Table 6-1. (*Continued*)

Outcome: Engaged community with diverse representation					
	X	X	X		Encourage and facilitate local community engagement throughout the project with representatives from all sectors and populations within the community

Preplanning	Planning	Design	Construction	Operation	Diversity
Outcome: A diverse and integrated community					
	X				Incorporate mix-use into the design
	X				Provide a range of housing types to reflect local diversity
		X			Design in flexibility of use and adaptability to change
Outcome: A diverse project team					
X	X	X	X	X	Ensure expertise on social and economic consequences and impacts are represented on the project team, either through explicit incorporation of team members, or through feedback from community engagement

Preplanning	Planning	Design	Construction	Operation	Mobility and Accessibility
Outcome: Connected communities					
X	X	X			Provide public transport options to connect communities and increase access to resources.
	X	X			Provide cycle lanes and footpaths
	X	X			Design for people not cars
	X				Provide walkable streets

Preplanning	Planning	Design	Construction	Operation	Safety
Outcome: Safe communities					
			X	X	Post clear and visible signs to indicate dangerous areas and egress areas
			X	X	Provide adequate outdoor lighting
		X			Incorporate secure by design and CPTED security design principles

Preplanning	Planning	Design	Construction	Operation	Economy
Outcome: Thriving local businesses					
X	X				Include local business retention as a required outcome
	X				Provide affordable workspaces for existing and local businesses
Outcome: Job opportunities for local people					
X				X	Require that job opportunities be advertised locally and be filled with local qualified applicants
	X	X	X	X	Provide a range of traineeships at difference levels and across difference fields
X	X	X	X	X	Provide work experience for local schools, especially in STEM subjects
			X		Organize site visits for local schools during construction
X	X				Engage with local businesses to identify how their services could be incorporated into the development process
X		X	X	X	Identify work packages that can be delivered by local Subject Matter Experts (SME)
Outcome: Increase productivity					
X	X				Reduce commute times by incorporating alternate transport links
Outcome: Economic diversity					
X	X				Encourage mixed use developments
X	X	X			Create an exciting space for incubating start-ups and new businesses

Source: Adapted from UKGBC (2018).

valuations depending on the community and its goals. Also, the dashboard concept could be used in design, as provided in Chapter 5. To ensure that a variety of metrics are used, it is important that a selection procedure be devised that allows tailoring of metrics to specific projects and locales. In developing the procedure, it is critical to understand and most important for all parties to agree that *one-size does not fit all*. As such, metric adoption and valuation vary depending on exposure, vulnerability, and local conditions.

6.3 CASE STUDIES

Three case studies are presented in this section to illustrate the thought process of accounting for key socioeconomic considerations in infrastructure project planning and design. Recall, the purpose of the metrics is to quantify the impact that an infrastructure project has on the surrounding community, specifically its social and economic institutions. Given that the proposed metric categories and subsequent goals are mostly newly proposed here and have not historically been incorporated into infrastructure design, a benchmark project does not exist to fully demonstrate using social and economic metrics here. The first case study adopts a hypothetical example of relocating a train station explaining the thought process behind incorporating six socioeconomic goals but does not fully demonstrate metrics and quantification. For quantification and application of such metrics, the reader is referred to Field et al. (2017). The second and third case studies adopt real-world examples to demonstrate how real projects have previously adopted socioeconomic considerations.

6.3.1 Simplified Hypothetical Case Study: Urban Train Station in Anywhere, United States

Consider a typical metropolitan city in the United States of approximately 500,000 to 1 million people, including suburbs, that is interested in shifting its major train station from the edge of the city to a more centralized location. The city is looking to set out a plan and subsequent design that addresses the following six socioeconomic needs of the community. Following each goal is an explanation of how the train station relocation project can meet the community goal and the associated metrics, in this case, in the wake of a major disaster.

6.3.1.1 Economic 1: Include Local Business Retention as a Required Outcome. Locating a large urban train station typically requires relocating local businesses (or other) that can be done through a number of mechanisms. In Anywhere, US, a plan is developed that allows local businesses to be relocated to an underground shopping center that is partially located under the train station and possesses all the technological advancements to attract customers such as restaurants, pubs, high-end and other shopping, movie theatre, bowling alley, laser tag, and so on.

Metrics: Number of local businesses (before and after development) and continuity of local businesses (before and after development). Quantifiable metrics include increases in local tax revenue, changes in household income, and employment rate. The values of these metrics should be set by the community because they are specific to the needs and goals within that community.

6.3.1.2 Economic 2: Provide Affordable Workspaces for Existing and Local Businesses. This development provides business hub spaces that rent space to those in transit, as well as local businesses at a reduced rate. Affordable rental space supports a thriving local economy, helps maintain the culture and character of the neighborhood, and gets buy-in from the local community.

Metrics: Number of affordable workspaces for existing and local businesses, number of local businesses (before and after development), and continuity of local businesses (before and after development).

6.3.1.3 Social 1: Provide Public Transport Options to Connect Communities and Increase Access to Resources. This new rail station connects local communities and provides easier access to jobs and other services. In addition, it incorporates community amenities that include a gymnasium, restaurants, a fitness/health center, grocery stores, a cinema, and green space with seating and benches.

Metrics: Reduction in average commute time, percentage of people within walking distance of resources, percentage of people using public transportation, increase in access to education, medical, and other facilities that do not exist in the local area, community services (recreational facilities, parks, historic sites, libraries, and museums) per 1,000 population, community walkability score, and reduction in transportation induced carbon emissions.

6.3.1.4 Social 2: Design for People Not Cars. Neighborhoods are too often designed for automobiles rather than people. This rail station focuses on people by providing safe access by foot and bicycle. The block around the station has been pedestrianized with a green space incorporating benches and safe bike paths. This improves community health and well-being by encouraging people to live healthier lifestyles and choosing to walk or bike rather than drive. The space is well maintained and includes a pleasant environment with good lighting and visibility. It incorporates the principles of (crime prevention through environmental design) (CPTED) to minimize criminal behaviors (see Gehl Architects "Human Scale" documentary 2020).

Metrics: Number of cars per 1,000 population, walkability score, air quality, percentage of green space per square kilometer, mileage of cycle lanes, reduction in available parking spaces, reduction in noncommunicable diseases, low crime rates, and reduction in transportation induced carbon emissions.

6.3.1.5 Social 3: Align Development with Community Values. Prior to starting the project, the developers held workshops with the local community to understand what they cared about (valued) as a community and what resources they would like to see incorporated into the rail station development. These workshops were intentionally located in different areas in the community and advertised across different media to recruit and engage diverse subsets of the population. This resulted in minimal objections and concerns that could have delayed the project later during construction and were addressed and resolved early in the process to all participants' satisfaction. The community benefits were measured before and after the project. This resulted in reputational benefits for the project developer and savings in financial costs because of the smooth planning process.

Metrics: Measure benefits against community values, increased community trust, and diverse and integrated community.

6.3.1.6 Social 4: Gather Community Feedback on the Development. To achieve Social 3, community engagement and feedback were essential. Empowering local residents to influence and take ownership of their local project made the project more successful. This was achieved through

1. Joint efforts to produce a mutually valued outcome,
2. Workshops held at different locations in the community,
3. Publicity campaigns and media engagement to maximize reach, and
4. Grassroots initiatives.

Metrics: Percentage of local support, percentage of positive feedback, diverse feedback, and diverse and integrated community.

Reflection: Through this hypothetical case study, it is clear that adoption of social and economic metrics into infrastructure projects, on the front end and through the project timeline can induce changes to engineering and design. For example, Social 1 would require adding train stops where there might otherwise not be stops to increase equitable accessibility. Similarly, Social 2 and Social 4 would require differences in the site plan, additional lighting, intentionally designed lines-of-sight, and the inclusion of additional amenities, such as bike racks. Finally, the proposed workshops in Social 3 might require changes to the typical preconstruction timeline and scheduling, as well as the role of the engineer, who may need to participate in these workshops to learn from community feedback. Overall, metric valuations (e.g., the actual percent limit specified in the metrics) should be set by the collaborative teams and based on community-set priorities.

6.3.2 Real Case Study: New York's Response to Hurricane Sandy

Hurricane Sandy, which struck New York City in 2012, was the largest ever recorded Atlantic hurricane and, after Katrina, the second most costly, causing damage of around US$70 billion in the United States alone (BS 67000) (BSI 2019). Hundreds of people were killed and hundreds of thousands were made homeless along the storm's path through the Caribbean, the United States, and Canada. However, although 24 US states were affected, it was the inundation of Lower Manhattan that generated the largest shock waves. The death, destruction, and general havoc wreaked by Sandy laid bare the inadequacies of current approaches to coastal flood risk management, generating a storm of public outrage.

The low-lying topography of Lower Manhattan from West 57th Street down to The Battery and up to East 42nd Street is home to approximately 220,000 residents and is the core of a $500 billion business sector that influences the world's economy. Hurricane Sandy devastated not only the Financial District but 95,000 low-income, elderly, and disabled city residents. Infrastructure within the 10 mi (16.0 km) perimeter was damaged or destroyed, transportation and communication were cut off, and thousands sat without power or running water.

The lesson from Sandy and New York City's response is that although there are good reasons why huge population centers have developed adjacent to and just a few feet above the ocean, living there involves flood risk; a risk that cannot be eliminated but can, and must, be reduced to a level that is acceptable or at least tolerable. To understand the limits of acceptable and tolerable, setting community goals established on engagement across all community sectors and walks of life is critical.

In collaboration with New York City, the BIG U proposal was developed to protect Lower Manhattan from floodwater, storms, and other impacts of a changing climate (Rebuild by Design 2021), as provided in Figure 6-2. The BIG U calls for a protective system around the low-lying topography of Manhattan beginning at West 57th Street, going down to The Battery, and then back up to East 42nd Street.

Figure 6-2. New York's Big U Project.
Source: Rebuild by Design (2021).

The proposed project provides 10 continuous miles of protection tailored to respond to individual neighborhood typology, as well as community-desired amenities. The proposal breaks the area into compartments: East River Park, Two Bridges and Chinatown, and Brooklyn Bridge to The Battery. Each compartment comprises a physically separate flood-protection zone, isolated from flooding in the other zones, but each zone offers about equally integrated social and community planning, thereby enhancing community engagement and cohesion. The compartments work in concert to protect and enhance the city, but each compartment's proposal is designed to stand on its own. Designs included paths and cycleways, children's play areas, and recreation space. This is a good example of providing an integrated solution with multiple benefits that includes health and well-being, community cohesion, and protection and environmental outcomes.

Metrics: Percentage of area green space per square kilometer, number of allotments/community gardens per square kilometer, and number of children's play areas per 1,000 population, ecological diversity of green space, percentage of reduction in carbon, number of trees planted, and percentage of community participation.

6.3.3 Real Case Study: City of Trees, City Re-Leaf Project, Manchester, United Kingdom

This project goes a step further than just metrics: It provides an example of quantifying the benefits of social and natural value in economic terms.

Reference sources are provided in the tables in the following. This illustrates the social, economic, and environmental benefits gained by designing green infrastructure.

The scope for creating large green spaces in the city centers is limited; however, street trees offer the opportunity to breathe life into gray streets, turning them into vibrant, livable spaces and places. They also help to combat climate change, improve our health and well-being, reduce surface water run-off, and remove pollution from air.

City of Trees has identified more than 1,000 locations across Manchester and Salford City centers in the United Kingdom where high-impact street trees could be planted (as illustrated in Figure 6-3).

The first trees were planted in 2019, and City of Trees plan to work with a range of partners to releaf the city center over the next 10 years. This ambitious program transforms the urban environment for the benefit of local business, people, and the environment.

Arup carried out a total value assessment of the benefits that City Re-Leaf will deliver for a range of stakeholders and beneficiaries in Manchester. Total value (Arup 2017) is an integrated approach using a combination of social value and natural capital methods to quantify and monetize the potential return on investment of City Re-Leaf. This differentiates the assessment from other discipline-bound studies that focus on one kind of value stream in isolation.

Figure 6-3. City of Trees, City Re-Leaf Project, Manchester, UK.

The assessment assumes that 100 trees will be planted each year for 10 years. The benefits have been prospectively assessed over a 50 year period from first planting, including land value, health and well-being, productivity, local spending, crime levels, carbon, and air quality, heat, and water. The study helps to make the wider business case, or the Total value case, for City Re-Leaf showing the broad range of benefits that can be achieved and captured through urban greening initiatives.

6.3.3.1 Metrics. The findings from the total value study are shown in Table 6-2. Over a 50 year period, the return on investment to local business, people, and the environment in the region will be £229 for every £1 spent on City Re-Leaf (2019 UK pounds, 1 pound = 1.2772 US dollars in 2019).

6.3.3.2 Evidence. The study findings were based on the following evidence:

1. Land value increases of 5% on tree-lined streets (Wolf 2007),
2. Improved mental well-being of 26% (Arup 2017),

Table 6-2. City Leaf Social Return on Investment.

Outcomes	Ratio	Total Value
Commercial land value	£1:£3	£22.5 million
Residential land value	£1:£0.36	£2.5 million
Well-being of residents	£1:£150	£1 billion
Productivity of workers	£1:£69	£482.3 million
Local spending	£1:£6	£42.2 million
Crime levels	£1:£0.01	£99.4 thousand
Carbon storage	£1:£0.01	£52.9 thousand
Air quality (NO_x, SO_x, and PM)	£1:£0.41	£2.9 million
Climate resilience (surface water management and urban cooling)	£1:£0.08	£0.6 billion
Total	£1:£229	£1.6 billion

Source: Arup (2017).

Note: In 2019 UK pounds. NO_x = nitrogen oxides; SO_x = sulfur oxides; and PM = particulate matter.

3. Increased productivity of desk-workers with a view of nature—estimate of 1% increase (Kaplan 1993),
4. Consumers willing to pay 9% to 12% more for goods and services with street trees (Wolf 2005),
5. Crime levels reduced by 12% for every 10% increase in tree cover (Troy et al. 2012),
6. Carbon and air quality benefits of street trees calculated using McPherson et al. (2002),
7. Water intercepted at a rate of 5.5 m^3 per tree (Song et al. 2018), and
8. Average heating/cooling cost savings of $32 per street tree in New York City (New York City 2015, Street Tree Census)

Valuation methods were based on the following:

1. Land value: Average commercial and residential rent prices in Manchester City Center.
2. Well-being: Housing Associations' Charitable Trust (HACT) Mental Well-being method.
3. Productivity: Gross value added income per head in Manchester City Center.
4. Local spending: Average spend per visitor in Manchester City Center.
5. Crime levels: HACT Social Value Bank method.
6. Carbon: UK Department for Business, Energy and Industrial Strategy traded carbon value for 2019.
7. Air quality: UK Department for Environment, Food and Rural Affairs 2015 damage costs (transport).
8. Water management (Endreny et al. 2017).

6.3.4 Real Case Study: Comprehensive Urban Resilience Masterplan for the City of Beirut, Lebanon

This project utilized a holistic resilience framework considering governance and economy, environment and infrastructure, and society community. This was overlaid with a risk-based assessment and qualitative stakeholder interviews to understand and prioritize actions for building resilience in Beirut, Lebanon.

Beirut is the backbone of the Lebanese economy. The city and the greater Beirut area are home to nearly half of the country's population and most of the financial, tourism, service, trade, government, educational, and health headquarters. Despite recurrent political, economic, and social shocks—both internal and external—Beirut continues to grow, overcoming multiple crises and fostering strong and vibrant social, cultural, and commercial sectors. Beirut faces an increasing spectrum and severity of risks, ranging from the potential for severe earthquakes and subsequent tsunami threats, and many other challenges such as climate change, mass migration, urbanization, aging populations, epidemics, poor air quality, and general infrastructure failure. The Comprehensive Urban Resilience Masterplan for the City of Beirut (World Bank 2017) provides the roadmap needed to make the city more resilient to these current and future challenges.

The project comprised a significant data collection and stakeholder engagement phase, a resilience diagnostic assessment of the city, followed by a resilience strategy and prioritized action plan. The resilience diagnostic included evaluation and prioritization of key shocks and stresses the city faces, mapping of the city assets and system vulnerabilities including demographics and socioeconomics, and mapping of future trends effects such as climate change and demographic change.

Owing to lack of data, house prices were used as a proxy for social inequality in this study, as shown in Figure 6-4.

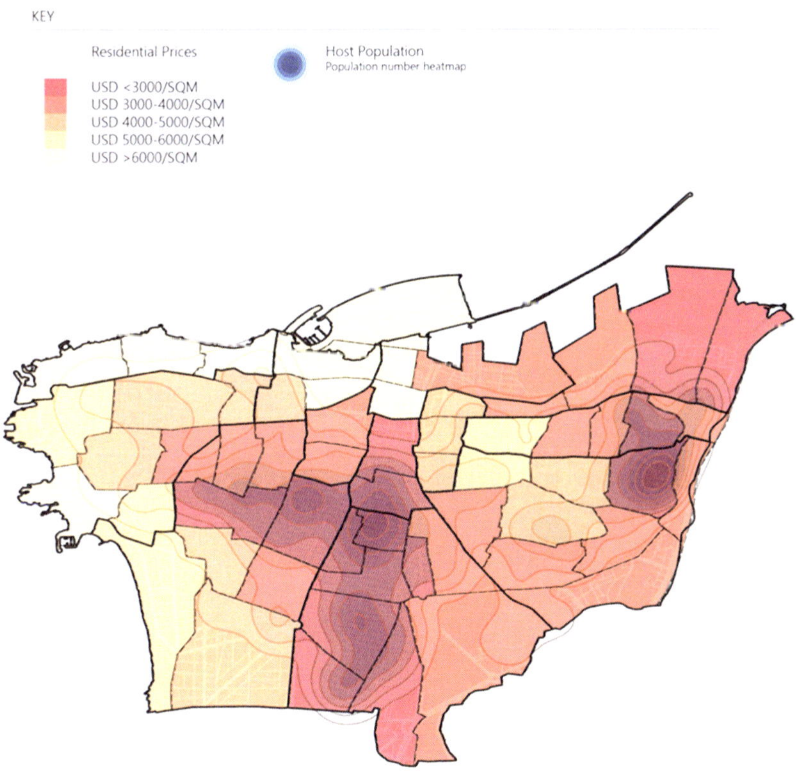

Figure 6-4. GIS map showing residential prices for Beirut.
Source: Courtesy of BuroHappold Engineering.

BuroHappold's resilience framework (Field et al. 2017) shown in Figure 6-5 was used to measure the resilience demand, capacity, and gaps across societal, governance, economic, and infrastructure sectors. The framework evaluates the risk to the systems owing to shocks and stresses and the capacity of these systems to mitigate or adapt. In addition, a stakeholder survey was performed to provide additional qualitative data. The questions around *community and inclusion* are shown in Table 6-3. The respondents were asked to strongly agree or disagree on a 5-point scale.

This provided the priority areas for the resilience strategy and action plan and also provided a benchmark for measuring the resilience benefit of the proposed actions. Actions included the establishment of a cross-sectoral platform for disaster risk management as a crucial aspect of risk governance to manage and implement the actions identified, an earthquake risk reduction plan, a climate change adaptation plan, a coastal resilience plan, an information management plan, a community resilience building plan, a response plan, and a recovery plan framework.

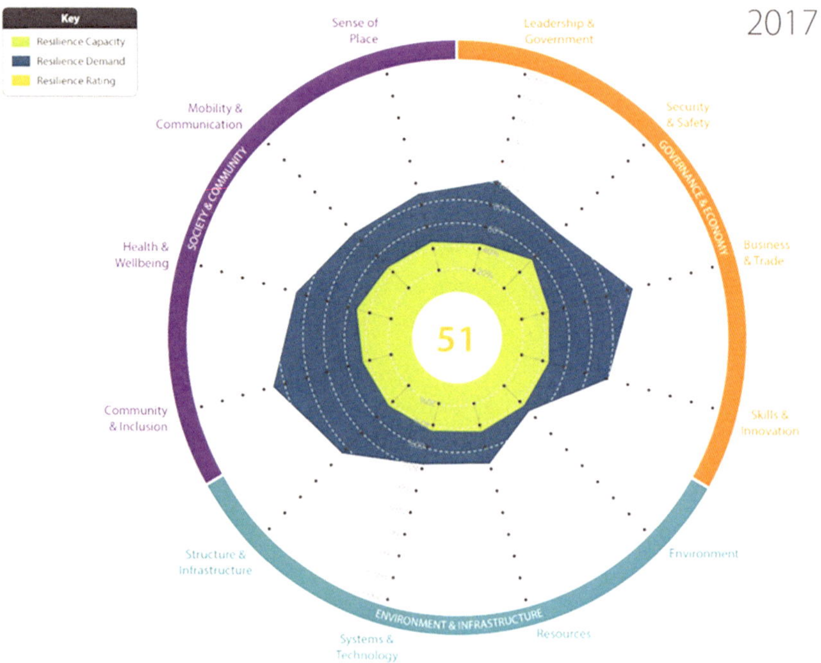

Figure 6-5. BuroHappold's City Resilience Framework. Source: Field et al. (2017).

Table 6-3. Excerpt of Questions from Beirut City Resilience Questionnaire.

Resilience Framework Category	Subcategory	Question
Community and inclusion	Social structure	Communities are integrated across the city with no segregation on social, economic, legal, or religious grounds
Community and inclusion	Community congregation	There are active centers for community congregation, for example, libraries, parks, and community centres, and regular community events are held
Community and inclusion	Community cohesion	Citizens have a strong sense of community; well-integrated, proactive, and mutually supporting
Community and inclusion	Voluntary sector	City has a strong voluntary sector with high percentage of the population volunteering
Community and inclusion	Role of voluntary sector in emergencies	Voluntary groups are considered an intrinsic part of emergency response
Community and inclusion	Community coordination	Community groups within the city have good coordination and governance
Community and inclusion	Community resilience program	Citizens and communities are well aware of their vulnerabilities and the hazards that put them at risk
Community and inclusion	Communication materials	Education and awareness materials and messaging promoting resilience are universally available in versions specific to community needs, for example, language
Community and inclusion	Resilience training	There is universal access to resilience related training, for example, first aid
Community and inclusion	Testing and exercising of the community	Community resilience plans are robust and regularly drilled/tested against a range of scenarios
Community and inclusion	Social protection	There is universal access to social protection mechanisms (welfare/benefit/social support)
Community and inclusion	Income inequality	There is very little deprivation, poverty, and income inequality within the city

Source: Courtesy of BuroHappold Engineering.

REFERENCES

Aldrich, D. P., and M. A. Meyer. 2015. "Social capital and community resilience." *Am. Behav. Sci.* 59 (2): 254–269.

Arup. 2017. *Fitzpark monitoring report*. London. Accessed January 6, 2020. https://www.arup.com/perspectives/publications/research/section/making-the-total-value-case-for-investment-in-infrastructure-and-the-built-environment.

BSI (British Standards Institute). 2019. *British standard for city resilience*. BS 67000. London: BSI. Accessed January 6, 2020. https://www.rebuildbydesign.org.

Burby, R. J., R. E. Deyle, D. R. Godschalk, and R. B. Olshansky. 2000. "Creating hazard resilient communities through land-use planning." *Nat. Hazard. Rev.* 1 (2): 99–106.

Chakalian, P., L. Kurtz, and D. Hondula. 2018. *Understanding vulnerability and adaptive capacity to large-scale power failure in the United States*. Quick Response Rep. QR 275. Broomfield, CO: Natural Hazards Center.

Chang, S. E. 2016. "Socioeconomic impacts of infrastructure disruptions." *Oxford Research Encyclopedia of Natural Hazard Science*. Accessed January 6, 2020. https://naturalhazardscience.oxfordre.com/view/10.1093/acrefore/9780199389407.001.0001/acrefore-9780199389407-e-66.

Chapin, F. S., III, S. R. Carpenter, G. P. Kofinas, C. Folke, N. Abel, W. C. Clark, et al. 2010. "Ecosystem stewardship: Sustainability strategies for a rapidly changing planet." *Trends Ecol. Evol.* 25: 241–249.

Cimellaro, G. P., C. Renschler, A. M. Reinhorn, and L. Arendt. 2016. "PEOPLES: A framework for evaluating resilience." *J. Struct. Eng.* 142 (10): 04016063.

Endreny, T., R. Santagata, A. Perna, C. De Stefano, R. F. Rallo, and S. Ulgiati. 2017. "Implementing and managing urban forests: A much needed conservation strategy to increase ecosystem services and urban wellbeing." *Ecol. Modell.* 360: 328–335.

Field, C., R. Look, and T. Lindsay. 2017. "A comprehensive approach to building and city resilience." In *Proc., AEI Convention.*, Oklahoma City, Oklahoma, April 11–13, 2017.

Fothergill, A., and L. A. Peek. 2004. "Poverty and disasters in the United States: A review of recent sociological findings." *Nat. Hazard.* 32 (1): 89–110.

Gehl Architects. 2020. "Human Scale, documentary." Accessed January 6, 2020. www.thehumanscale.dk.

Hamideh, S., and J. Rongerude. 2018. "Social vulnerability and participation in disaster recovery decisions: Public housing in Galveston after Hurricane Ike." *Nat. Hazard.* 93 (3): 1629–1648.

Hendricks, M. D. 2017. "The infrastructures of equity and environmental justice." *Ph.D. thesis*, Texas A & M University, Dept. of Civil and

Environmental Engineering. Accessed January 6, 2020. http://hdl.handle.net/1969.1/161342.

Hendricks, M. D., M. A. Meyer, N. G. Gharaibeh, S. Van Zandt, J. Masterson, J. T. Cooper, et al. 2018. "The development of a participatory assessment technique for infrastructure: Neighborhood-level monitoring towards sustainable infrastructure systems." *Sustainable Cities Soc.* 38: 265–274.

Hondula, D. M., R. C. Balling, J. K. Vanos, and M. Georgescu. 2015. "Rising temperatures, human health, and the role of adaptation." *Curr. Clim. Change Rep.* 1: 144–154.

Johansen, C., J. Horney, and I. Tien. 2017. "Metrics for evaluating and improving community resilience." *J. Infrastruct. Syst.* 23 (2): 04016032.

Kaplan, R. 1993. "The role of nature in the context of the workplace." *Landscape Urban Plann.* 26 (1–4): 193–201.

Koliou, M., J. W. van de Lindt, T. P. McAllister, B. R. Ellingwood, M. Dillard, and H. Cutler. 2020. "State of the research in community resilience: Progress and challenges." *Sust. Resil. Infrastruct.* 5 (3): 131–151.

Kruchten, P., C. Woo, K. Monu, and M. Sotoodeh. 2007. "A human-centered conceptual model of disasters affecting critical infrastructures." In *Proc., Intelligent Human–Computer Systems for Crisis Response and Management*, edited by B. Van de Walle, P. Burghardt, and C. Nieuwenhuis.

Levine, J. N., A. M. Esnard, and A. Sapat. 2007. "Population displacement and housing dilemmas due to catastrophic disasters." *J. Plann. Lit.* 22 (1): 3–15.

McPherson, E. G., S. E. Maco, J. R. Simpson, P. J. Peper, Q. Xiao, A. VanDerZanden, et al. 2002. *Western Washington and Oregon community tree guide: Benefits, costs and strategic planting*. Accessed March 11, 2020. https://www.fs.usda.gov/treesearch/pubs/45962.

Meyer, M. A., M. Hendricks, G. D. Newman, J. H. Masterson, J. T. Cooper, G. Sansom, et al. 2018. "Participatory action research: Tools for disaster resilience education." *Int. J. Disast. Resil. Built Environ.* 9 (4/5). 402–419.

Mitsova, D., A.-M. Esnard, A. Sapat, and B. S. Lai. 2018. "Socioeconomic vulnerability and electric power restoration timelines in Florida: The case of Hurricane Irma." *Nat. Hazard.* 94 (2): 689–709.

Multihazard Mitigation Council. 2018. *Natural hazard mitigation saves: 2018 Interim report*. Washington, DC: National Institute of Building Sciences. Accessed January 6, 2020. www.nibs.org.

New York City. 2015. "Street tree census report." Accessed March 10, 2020. http://media.nycgovparks.org/images/web/TreesCount/Index.html.

NIST (National Institute of Standards and Technology). 2015. *Community resilience planning guide for buildings and infrastructure systems*, Vol. 1. NIST Special Publication 1190. Washington, DC: US Dept. of Commerce.

Peacock, W. G., S. Van Zandt, Y. Zhang, and W. E. Highfield. 2014. "Inequities in long-term housing recovery after disasters." *J. Am. Plann. Assoc.* 80: 356–371.

Ritchie, L. A., and D. A. Gill. 2011. *Vol. 14 of Considering community capitals in disaster recovery and resilience*. PERI Scope. Fairfax, VA: Public Entity Risk Institute. Accessed January 6, 2020. https://www.riskinstitute.org/peri/component/option,com_deeppockets/task,catContShow/cat,86/id,1086/Itemid,84/.

Shirley, W. L., B. J. Boruff, and S. L. Cutter. 2012. "Social vulnerability to environmental hazards." In *Hazards vulnerability and environmental justice*, edited by S. Cutter, 143–160. Abington, UK: Routledge.

Song, X. P., P. Y. Tan, P. Edwards, and D. Richards. 2018. "The economic benefits and costs of trees in urban forest stewardship: A systematic review." *Urban For. Urban Greening* 29: 162–170.

Sutley, E. J. 2018. *Ending bias in disaster mitigation and recovery policies*. Research Counts. Boulder, CO: Natural Hazards Center. Accessed January 6, 2020. https://hazards.colorado.edu/news/research-counts/ending-bias-in-disaster-mitigation-and-recovery-policies.

Sutley, E. J., J. W. van de Lindt, and L. Peek. 2016a. "Community level framework for seismic resiliency I: Coupling socioeconomic characteristics and engineering building systems." *Nat. Hazard. Rev.* 18(3), https://doi.org/10.1061/(ASCE)NH.1527-6996.0000239

Sutley, E. J., J. W. van de Lindt, and L. Peek. 2016b. "Community level framework for seismic resiliency II: Multi-objective optimization and illustrative examples." *Nat. Hazard. Rev.*

Sutley, E. J., J. W. van de Lindt, and L. Peek. 2017. "Multihazard analysis: Integrated engineering and social science approach." *J. Struct. Eng.* 143 (9), 143(9), https://doi.org/10.1061/(ASCE)ST.1943-541X.0001846.

Troy, A., J. M. Grove, and J. O'Neil-Dunne. 2012. "The relationship between tree canopy and crime rates across an urban–rural gradient in the greater Baltimore region." *Landscape Urban Plann.* 106 (3): 262–270.

UKGBC (UK Green Building Council). 2018. "Social value in new development: An introductory guide for local authorities and development teams." Accessed July 20, 2020. https://www.ukgbc.org/wp-content/uploads/2018/03/Social-Value.pdf.

van de Lindt, J. W., W. G. Peacock, J. Mitrani-Reiser, N. Rosenheim, D. Deniz, M. Dillard, et al. 2018. "Community resilience-focused technical investigation of the 2016 Lumberton, North Carolina Flood: Multi-disciplinary approach." Edited by J. W. van de Lindt, W. G. Peacock, and J. Mitrani-Reiser. *NIST Special Publication 1230*. Gaithersburg, MD: National Institute of Standards and Technology.

van de Lindt, J. W., W. G. Peacock, J. Mitrani-Reiser, N. Rosenheim, D. Deniz, M. Dillard, et al. 2020. "Community resilience-focused technical investigation of the 2016 Lumberton, North Carolina Flood: An interdisciplinary approach." *Nat. Hazard. Rev.* 21 (3): 04020029.

Wolf, K. L. 2007. "City trees and property values." *Arborist News* 16 (4): 34–36.

Wolf, K. L. 2005. "Business district streetscapes, trees, and consumer response." *J. For.* 103 (8): 396–400.

World Bank Group, BuroHappold Engineering. 2017. "Comprehensive urban resilience master plan for the city of Beirut." Accessed January 6, 2020. https://www.gfdrr.org/en/publication/results-resilience-comprehensive-urban-resilience-master-plan-beirut.

Rebuild by Design. 2021. New York's Big U project, Accessed March 12, 2021. http://www.rebuildbydesign.org/our-work/all-proposals/winning-projects/big-u.

CHAPTER 7
EMERGING RESILIENCE-ENABLING TECHNOLOGIES

7.1 INTRODUCTION

Resiliency of existing and new infrastructure has become a constant theme in government and industry discussions. To operate optimally and to ensure system safety and resilience of critical infrastructure, there is a need to deploy innovative construction materials, utilize smart methods in construction technology, and deploy sensor networks and monitoring technology in infrastructure systems. On the one hand, emerging new materials and construction technologies are expected to contribute to improving resilience capacities for our infrastructure, namely, absorptive, adaptive, and restorative capacities. The distinct material properties and structural features of new construction materials and methods enable infrastructure to withstand or efficiently recover from multihazard disruptive events leading to improved resilience and reduced life-cycle cost. On the other hand, using recent advancements in sensing technologies and communication networks from the internet of things (IoT) to the rich information obtained from sensors embedded within the infrastructure allows us to develop a new paradigm in designing, constructing, operating, and maintaining infrastructure systems. The potential value of sensing is not only to understand the behavior of monitored structures but also to make better decisions for asset management of infrastructure at the city scale, enabling life-cycle-based thinking. We argue that the integration of three distinct disruptive technologies results in altering design methods for resilient infrastructure. These technologies are smart materials, advanced construction technologies, and advanced sensing technologies, as depicted in Figure 7-1. The figure also provides an insight into the relationship between the emerging technologies and the four characteristic elements

Figure 7-1. Schematic representation of the integration of emerging technologies and their relation to the four characteristic elements of resilient infrastructure.

(known as 4Rs) of infrastructure resilience, namely, redundancy, robustness, rapidity, and resourcefulness. This chapter discusses these technologies in detail, as well as their potential impact on infrastructure resilience. It is important to note that these three classes of technology and their influence on infrastructure resilience strongly overlap; both smart materials and sensing technologies are used in construction. Today, numerous contractors worldwide deploy unmanned aerial systems (UASs) to monitor construction sites. Furthermore, smart materials include self-sensing technologies. It is not the objective of this book to create such classification but rather to describe the main categories of emerging technologies and examine their specific applications to resilient infrastructure.

7.2 ADVANCED AND SMART MATERIALS

7.2.1 Multifunctional Fiber and Polymer Composites

Because of its ease of use and high compressive strength, concrete is the most commonly used construction material worldwide. However, concrete's weak tensile strength represents a major limitation in the performance of concrete structures under extreme loading conditions (e.g., earthquakes). This limitation has resulted in oversized and heavily reinforced structural members, where ductility and strains govern the design (Ranade et al. 2017). Furthermore, all structural design codes neglect to reflect any tensile capacity contribution of concrete because of its brittleness.

In the 1950s, polymer concrete (PC) was introduced as the main construction material in the infrastructure sector (ACI 548.1R-09) (ACI 2009). Fiber-reinforced polymers (FRPs) were first introduced to aerospace in the 1960s, migrating to the automobile industry in the 1970s, and is now a well-established technology used in the construction industry (Mosallam 2011). Since the 1990s, major research has enabled the technological transfer of polymer composites from the research labs to the field.

In the last two decades, the introduction of polymers, fibers, and nanomaterials represents significant changes in the concrete industry. Recently, researchers have proposed new techniques to produce concrete that overcomes the weak tensile strength and brittleness of traditional concrete. Examples of concrete with improved tensile performance include engineered cementitious composites (ECCs), strain hardening cement composites (SHCC), and PC (Ranade et al. 2017, Emiroglu et al. 2017). These proposed concretes exhibit multicracking behavior, high tensile strength (>10 MPa), and high ductility (>3%). Ultrahigh-performance (UHP) ECCs (UHP-ECCs) have been developed by Yu et al. (2018) by incorporating high aspect ratio micro-polyethylene (PE) fibers achieving strains at failure up to 7.2% and a tensile strength of 17.4 MPa. Similar high-strength high-ductility concrete (HSHDC) has been developed by Ranade et al. (2017) using synthetic fibers with strain at failure of 3.4%, a tensile strength of 16 MPa, and improved impact resistance with higher damage tolerance. PC with alumina nanoparticles has been shown by Emiroglu et al. (2017) to reach strain at failure of 4.9% with tensile strength greater than 10 MPa and fracture toughness improvement greater than 100%. Figure 7-2 shows additional examples of PC subject to bending and direct tension, where the surface exhibits stretching, as well as multicracking

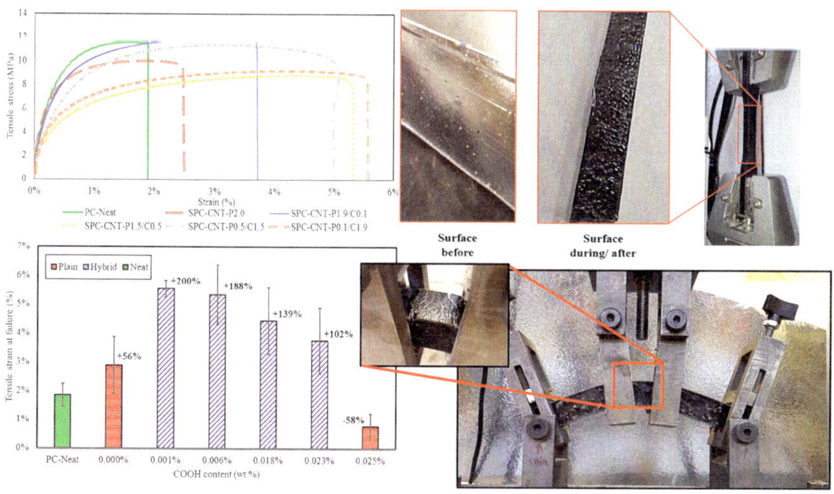

Figure 7-2. Multifunctional PC with extreme ductility and sensing capabilities incorporating nanomaterials.
Source: Emiroglu et al. (2017).

with tensile strains of up to 5.5%. The tensile properties of PC can be engineered by manipulating the ratio of functionalized and nonfunctionalized multiwalled carbon nanotubes (MWCNTs), as shown in Figure 7-2.

The recovery rate of strength and stiffness in self-healing materials is material-specific and highly depends on the self-healing technique employed. Self-healing materials recovery ranges from a few days to a few weeks. Self-healing materials with relatively fast recovery might influence the recovery period of a structure after a disruptive shock and thus can improve *rapidity* as a main component of infrastructure resilience.

Considerable progress has been made in reinforcing polymers with nanomaterials by incorporating organofunctional groups to improve covalent bonds throughout the interface (Namkanisorn et al. 2001, Abel et al. 2004, Berry and Namkanisorn 2005). The proposed group of polymer nanocomposite material includes epoxy, polyester, and vinyl ester polymers incorporating functionalized and nonfunctionalized MWCNTs. Furthermore, other nanomaterials were also reported to achieve excellent mechanical characteristics when incorporated in epoxy such as graphene nanoplatelets (GNPs) and boron nanotubes (BNTs) (Zhou et al. 2013).

Researchers have shown that significant improvement of shear strength, fracture toughness, and creep characteristics of epoxy can be achieved with the addition of carbon nanotubes (Soliman et al. 2012). The literature reports remarkable levels of ductility that are at an order of magnitude, or more, higher than portland cement concrete. The literature also reports numerous examples of what could potentially become self-sensing capabilities of cement, PC, and composites with the addition of nanomaterials (Wen and Chung 2002, Yu and Kwon 2009, Al-Sabagh et al. 2017, Ghafari et al. 2018). Such technology not only allows significant improvement in the mechanical performance of concrete and composites but also adds to the potential of multifunctionality to infrastructure materials defined by expanded and broad uses. The broad multifunctionality of concrete and composites can contribute to adaptive capacities necessary for resilient infrastructure in the future.

7.2.2 Textile-Reinforced Composites

Textile-reinforced concrete (TRC) has emerged in the last decade as a new and valuable construction material. TRC is a new type of a cementitious composite that has been promoted for the fast fabrication of new structural elements and for the strengthening of existing structures. TRC is made with a continuous textile fabric that is incorporated into a cementitious matrix consisting of a portland cement binder and small-sized aggregates (Lepenies et al. 2008, Peled et al. 2016). Example textile reinforcement used in TRC is shown in Figure 7-3. Reinforcement in TRC

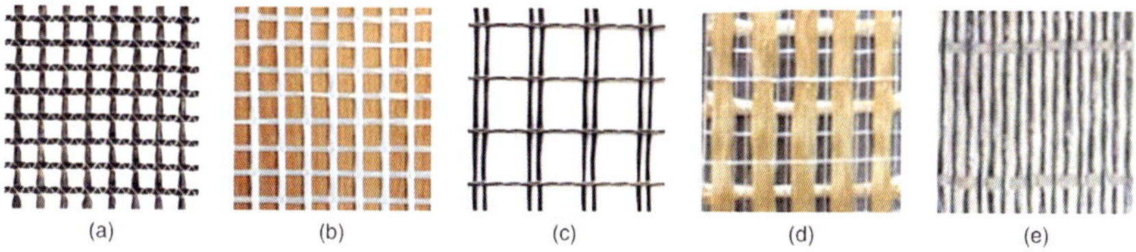

Figure 7-3. Examples of textile reinforcement: (a) carbon fibers, (b) glass fibers, (c) basalt fibers, (d) polyphenylene benzobisoxazole fibers, and (e) steel fibers. Source: Koutas et al. (2019).

is typically high-strength and high-modulus fibers and filaments (continuous fibers) that are made of basalt, polypropylene, alkali-resistant (AR) glass, carbon, and other polymers (Du et al. 2018, Volkova et al. 2016, Žák et al. 2017, Zastrau et al. 2003). Textile reinforcement may be produced either planar (2D) or spacer (3D). 3D reinforcing textile allows for resisting in-plane and out-of-plane loads (Gries et al. 2016).

In the last decade, TRC has been evaluated extensively at RWTH Aachen University and Dresden University of Technology. Researchers in the US, Brazil, and Israel have also examined TRC (Dalal et al. 2017, Orlowsky and Raupach 2011, Portal et al. 2014). TRC can be used to build slender, lightweight, modular, and free-form structures, eliminating the risk of corrosion. In addition, TRC is a suitable solution for the strengthening of existing structures (Williams Portal et al. 2015). Owing to the superior characteristics of the TRC, it has become a focus of interest for many. TRC has certain advantages; it is lightweight, resistant to high temperatures and corrosion, and it has high strength in compression and tension, superior ductility, crack width reduction, a high strength-to-weight ratio, ease of handling, quick installation, modular and free-form structures' design aspects, visual impact, and reversibility. TRC can be classified as a sustainable construction material because of these superior characteristics (Abdel-Emam et al. 2018, Dalal et al. 2017, Jamshaid et al. 2018, Volkova et al. 2016, Žák et al. 2017).

Applications of TRC in the construction industry extend to ventilated facade systems in small, mid-size, and large formats, sandwich walls, modules (e.g., garages and transformer stations), storage units (tanks, silos, and similar), bridges (new construction and maintenance), free-form surfaces, load-bearing shell structures, balcony slabs, building elements with high chloride exposure (e.g., parking garage ceilings), maritime building elements, load-bearing structure reinforcements, concrete remediation, sprayed concrete applications, and noise/water protection walls (Abdel-Emam et al. 2018, Kulas 2015). TRC represents an ideal example of lightweight flexible materials that can be used to design and construct thin, complex, and resilient infrastructure with superior ductility.

7.2.3 Superelastic Materials

Shape memory alloys (SMAs) are another class of advanced materials with the ability to *memorize* or retain their previous form when subjected to either thermally induced (shape memory effect SMAs) or stress-induced (superelastic SMAs) stimulus, as shown in Figure 7-4. The memory capabilities depend mainly on the SMAs' diffusionless solid-to-solid phase changes as metallic alloys. Nickel–titanium (NiTi)-based SMAs have been the pioneering alloy type to be utilized for commercial purposes. NiTi SMAs can recover from large strains between 6% and 8% and recover from large stresses of as much as 90% of yield strength (Alaneme and Okotete 2016). In addition, NiTi SMAs display high ductility with excellent corrosion resistance and fatigue behavior. However, full utilization of NiTi SMAs in commercial applications can still be limited by their high cost and complexity of production. Therefore, copper- and iron-based SMAs are recently viewed as potential cost-effective substitutes to NiTi SMAs based on their promising shape memory and other functional properties.

SMAs have drawn significant attention and interest in a broad range of industrial and numerous fields such as biomedical, aerospace, and automotive applications (Jani et al. 2014). In industrial systems, SMAs have been used as fluid connectors and couplers and as thermal actuators in fire alarms and fire safety valves.

In infrastructure, SMA elements are mostly used in the forms of wires, bars, and more recently prestressing strands. Superelastic SMA wires were implemented in bracing elements for tall structures (Ozbulut et al. 2016),

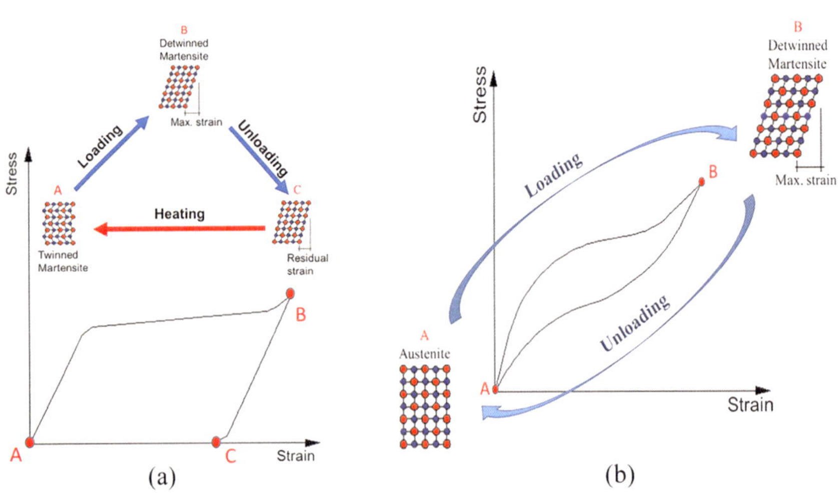

Figure 7-4. (a) Shape memory effect, and (b) superelastic effect.
Source: Ozbulut et al. (2016).

and they were also combined with natural rubber bearings and superelastic friction base isolators to produce SMA-based rubber bearing systems (Ozbulut and Hurlebaus 2012). In contrast, heat-activated prestrained martensite SMA wires were used as confining jackets in the plastic hinge regions of lap-spliced reinforced concrete (RC) columns for seismic retrofitting applications (Choi et al. 2012). More recently, superelastic SMA wires and small-diameter SMA strands were utilized as reinforcing fibers in epoxy matrices to produce FRP composites with high strain capabilities (Daghash and Ozbulut 2016, 2018), whereas thermally responsive SMA wires were used in elastomer matrices to develop multifunctional composites (Feng et al. 2015). Furthermore, superelastic SMA bars were incorporated as the main reinforcement in the plastic hinge regions of RC columns to reduce permanent displacement and limit damage in columns subjected to strong earthquakes (Tazarv and Saiidi 2015). They also replaced regular steel bars employed as longitudinal reinforcement in ductile fiber-reinforced cement-based composite beams providing self-centering and deflection reversal properties (Hung and Yen 2014). Finally, the cyclic and fatigue properties of SMA cables were studied. It was suggested to place SMA bars in high-capacity structural applications (Ozbulut et al. 2016, Sherif and Ozbulut 2018). SMAs' large recoverable strain capacity, along with its exceptional energy absorption capabilities, can play an important role in constructing the next generation of resilient infrastructures.

7.2.4 Self-Healing Materials

Autonomous detection, damage repair (i.e., self-healing), and efficient material repair monitoring are a powerful combination for improving the reliability and extending the service life of structural material systems. Lin et al. (1990) showed the potential healing of polymer cracks using methanol. Capsule-based approaches were first achieved by White et al. (2001), whereby incorporating monomer-filled capsules embed a catalyst within the matrix material damage repair that could be induced on microcapsule rupture. Subsequently, microcapsule-based approaches to self-healing have been studied extensively and applied to a range of systems (Aïssa et al. 2012, Blaiszik et al. 2010, Brown et al. 2002, Caruso et al. 2007, Duncanson et al. 2012, Jin et al. 2011, Yuan et al. 2008, Yang et al. 2013).

There are several techniques for forming microcapsules, including nanoprecipitation, in situ polymerization, emulsification, coacervation, and interfacial polymerization. Initial microcapsules are prepared with a poly(urethane)/urea–formaldehyde (PU–UF) shell using a method analogous to Gladman et al. (2015). Microcapsules with PU–UF shells serve as the benchmark for microcapsule property comparisons. This technique allows the encapsulation of hydrophobic species such as the monomer and nanoparticle species for self-healing analogous to the method pioneered by White et al. (2001), as well as the encapsulation of hydrophobic solvents that also induce self-healing through the so-called solvent healing (Gladman et al. 2015, Celestine et al. 2015). For solvent healing, a good solvent [a solvent similar in solubility parameter (δ) to the matrix polymer] for the matrix polymer is encapsulated and on rupture is released. On release, the surrounding polymer matrix swells, and in epoxy-based and polymethyl methacrylate-based systems, such as those of interest herein, this type of solvent swelling has been shown to lead to self-healing (Caruso et al. 2007, 2008; Celestine et al. 2015). For example, in an epoxy matrix material, the release of the solvent from the microcapsule swells the matrix, causing the cracks to narrow, and promotes additional cross-linking between residual amine groups bridging the crack surface, effectively healing the matrix.

Likewise, self-healing in polymer-based materials and cementitious materials gained similar attention because of the frequent use of concrete in construction and its high vulnerability to cracking under many service-loading and environmental conditions. Over the years, several strategies were introduced for self-healing cementitious materials, such as vascular-based healing, microencapsulation, expansive agents, bacterial repair, and shape memory materials (Wu et al. 2012, Van Tittelboom and De Belie 2013). In the first strategy, the healing agent is usually stored inside a hollow fiber network, also known as pipettes, and is released on cracking (Dry 1994, Mihashi et al. 2001). In the microencapsulation approach, the self-healing agent is stored in discrete capsules rather than hollow fibers and is released when capsules are ruptured. Healing agents harden when exposed to moisture, air, heat, and/or when they come in contact with the cementitious matrix itself (Cailleux and Pollet 2009, Boh and Šumiga 2008). Healing with expansive agents relies on adding expansive mineral admixtures that can react later with existing water when damage occurs to seal the cracks or lead to recrystallization of hydrated cement on the crack surface (Ahn and Kishi 2010). Bacterial repair involves microbial precipitation of calcium carbonate ($CaCO_3$) in the cracks using concrete-immobilized bacteria (Jonkers and Schlangen 2008, Jonkers et al. 2010). The use of shape memory materials (e.g., alloys or polymers), also known as intelligent reinforced concrete (IRC), provides mechanical self-healing techniques by reforming the shape memory materials to their original shape when heated or mechanically stressed, leading to crack closure (Kuang and Ou 2008). All the self-healing strategies for polymer- or cement-based materials can benefit the restorative capacity of infrastructure resilience by providing mechanisms of self-recovery following exposure to multihazard events. Self-healing of construction materials could also contribute to improving the durability of infrastructure and reducing the vulnerability of cracking in aggressive environments.

7.2.5 Bioinspired Materials

In addition to self-healing, nature inspires material scientists and engineers with many other efficient materials and structural features that can be of great benefit to the civil infrastructure and construction industry. Examples of different potential applications for bioinspired materials in the construction industry are shown in Figure 7-5. A similar example was provided by Pereira et al. (2015). The figure introduces a variety of features that can be implemented in the construction industry to build a more resilient infrastructure. Of great interest to structural engineering is the unique microstructure of many natural materials and structures. In particular, the multiscale hierarchical features from nano- to macroscales and complex arrangement of soft and hard materials provide distinguished structure–property relationships (i.e., combination of stiffness, strength, fracture, impact resistance, and low density) that can satisfy resiliency needs for civil infrastructure.

Examples of natural materials that attracted scientists include the microstructure structure of plant stems, bone, bamboo, and nacre. Stems in plants, for instance, represent a hierarchical structure with five structural levels: biochemical (chemical composition), ultrastructural (cell wall), microscopic (cell), macroscopic (tissue), and integral (stem) levels (Speck and Rowe 2006, Knippers and Speck 2012). Another interesting microstructure can be found in nacre, which consists of a dovetailed brick

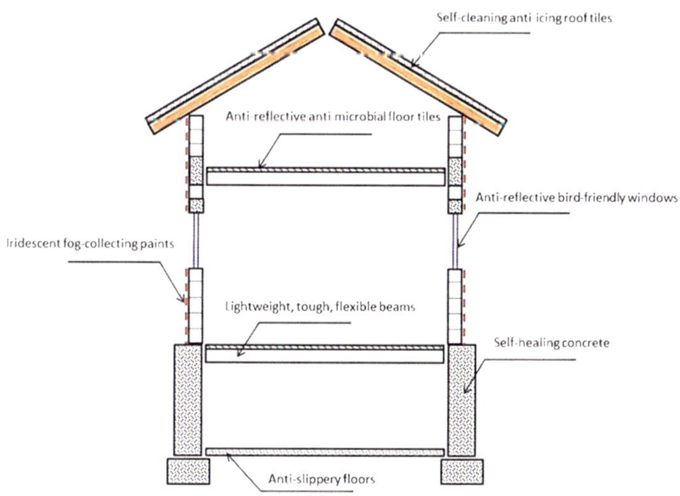

Figure 7-5. Schematic illustration of various potential applications of bioinspired features in the construction industry.
Source: Pereira et al. (2015).

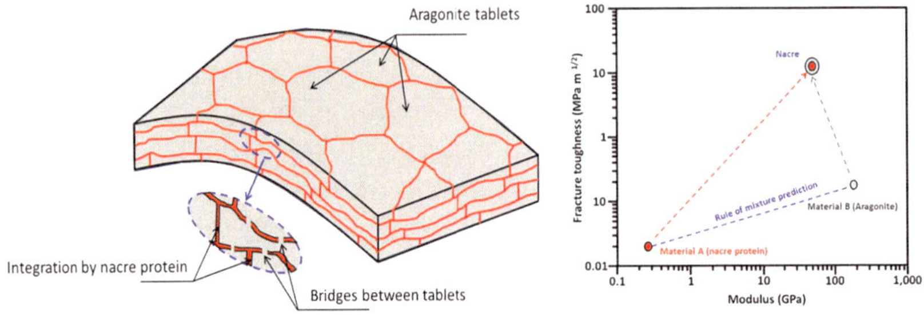

(a) (b)

Figure 7-6. Unique microstructure of nacre and its influence on the material properties: (a) aragonite tablet bridges at specific locations [schematic illustration after work by Sun et al. (2016)], and (b) material property chart showing improved fracture toughness with respect to modulus.
Source: Data are extracted from Wegst et al. (2015).

and mortar structure, as shown schematically in Figure 7-6(a) (Sun et al. 2016). The distinction in nacre and other natural materials and structures arises from combining several components with poor intrinsic properties to produce high-performance structures. Such a combination does not typically follow simple rules of mixtures, as shown in Figure 7-6(b) (Wegst et al. 2015).

The figure depicts a material property chart for several natural and synthetic materials. The figure demonstrates how combining natural materials with low intrinsic properties (e.g., nacre protein and aragonite or collagen and hydroxyapatite) in a unique microstructure could yield materials with superior properties (e.g., nacre and bone). Because of this unique microstructure, many attempts have been made to study and mimic the architecture of nacre. Sun et al. (2016) investigated the shear, compression, bending, and uniaxial tension load transfer and energy dissipation mechanisms in nacre and demonstrated how such a microstructure would benefit the design of impact resistance cement- and clay-based structures. Munch et al. (2008) utilized an ice-templated microstructure to produce extremely tough alumina/methylmethacrylate synthetic composite by mimicking the microstructure of nacre. Using such bioinspired materials can enhance the absorptive capacity of resilient infrastructure by improving robustness and providing unique energy absorption mechanisms for construction materials and structural components when subjected to extreme loading events leading to less disruption and more functionality.

At the structural level, several examples exist for infrastructure that mimic the architecture in nature. Early examples of bioinspired infrastructure include the Geodesic Dome in Montreal, constructed in 1967, aimed at developing a spherical topology that utilizes repeated components of similar beams and nodal joints following the typical structure of many natural structures. Since then, several other bioinspired architectural designs have been constructed, such as the Swiss Re Gherkin Tower in London, which mimics a type of marine organism, the Kunsthaus Graz in Australia, which mimics a Burgess Shale fossil, and the Hörn folding bridge in Germany, which mimics multifunctionality and adaptivity of natural systems (Aldersey-Williams 2004, Knippers and Speck 2012). Knippers and Speck (2012) highlighted several architectural features that can be incorporated in the design to produce biomimetic materials or structures, including heterogeneity, anisotropy, hierarchy, and multifunctionality. Although it is apparently challenging to incorporate many of these features in real construction, computer-aided design and new construction technologies, such as 3D printing technology, also known as additive manufacturing, have shown very promising capabilities to execute bioinspired designs, as discussed in the following.

7.3 ADVANCED CONSTRUCTION TECHNOLOGY

In parallel to the advancement of infrastructure materials, numerous developments are taking place in construction technologies, leading to the so-called Third Industrial Revolution. The recent advancement in the field of construction technologies stands on two distinct pillars. The first pillar is related to the extensive development in the field of information technology and networking. The growth in information technology enabled the large involvement of computers and networks in the construction field. The second pillar represents recent advancements in machinery techniques and capabilities, starting from computer numerical controlled (CNC) machines and extending to additive manufacturing technology. The two pillars combined pave the way for the development and growth of *digital fabrication* in the construction industry. The following sections briefly cover two important technologies that are developed following these two directions (Naboni and Paoletti 2015).

7.3.1 Building Information Modeling

Building information modeling (BIM) is a comprehensive and collaborative framework that utilizes computer modeling to manage various steps in structural/architectural engineering design and construction. This integrated platform enables complete virtual design, construction, and occupation of structures and serves different professionals involved in the construction industry, including owners, design and construction engineers, managers, and contractors (Kensek and Noble 2014, Eastman et al. 2012). BIM was first introduced as a *building description system* in the mid-1970s by Eastman (1975) to cover several aspects of the current technology. Currently, BIM technology is being utilized in the design and construction of new buildings, as well as in the maintenance and demolition/deconstruction of existing buildings (Volk et al. 2014).

The increasing use of BIM is driven by the need for simplifying design, promoting efficient coordination and planning, ensuring safe construction, and achieving satisfactory maintenance. A typical BIM framework involves three overlapping fields: technology, process, and policy—known as TPP fields (Succar 2009). The first field involves the infrastructure (hardware and software) employed to execute various tasks. The process field involves various work activities (design, planning, construction, and manufacturing) carried out by professional personnel. The third field, policy, involves rules and guides that are followed to control the work progress and promote collaboration between various disciplines. One of the distinct features of BIM technology is that it also enables optimizing design parameters to achieve more robust design, as shown schematically in Figure 7-7. In the example shown, the BIM framework includes parametric modeling to establish relations between various objects used in virtual design and construction (Kensek and Noble 2014).

Although BIM technology seems very appealing, the framework requires a large availability of information and the establishment of complex networking. In addition, assessment of BIM performance is extensive and carried out on five different components, namely, BIM capabilities, maturity levels, competency sets, organizational sets, and granularity levels. These five components ensure evaluating the BIM framework from different perspectives (e.g., technicality, efficiency, size,

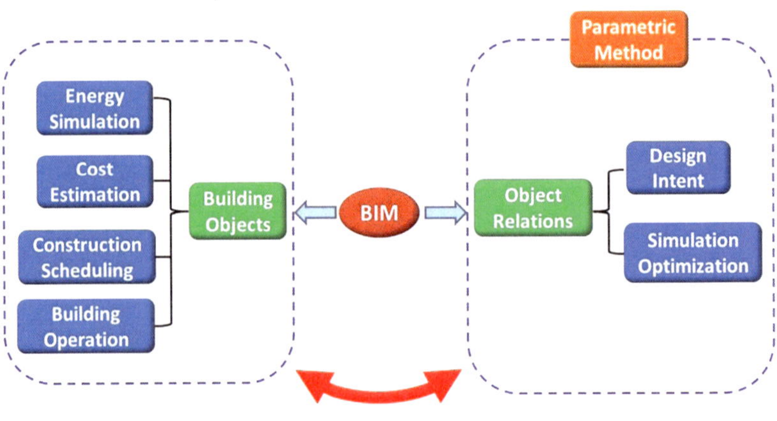

Figure 7-7. General BIM building objects associated with parametric modeling.

and documentation) (Succar 2010). The ability of BIM technology to provide a wealth of information on all structural and construction details provides notable resourcefulness that can enable fast recovery and thus will have a remarkable impact on infrastructure resilience.

7.3.2 Artificial Intelligence and Machine Learning

Artificial intelligence (AI) is a term used to describe the computational methods developed to simulate human cognition and enable artifacts to mimic human intelligence by having perception, reasoning, knowledge representation, planning, learning, language processing, and the ability to manipulate information or objects in their environments (Lugar 2009). The term, AI, was used for the first time in 1956 (Lu et al. 2012). Significant research was done in the last 40 years to enable the use of AI in civil engineering. Several good review articles were published in the last decade and traced the inception, development, and applications of AI in civil infrastructure (Adeli and Hung 1994, Adam and Smith 2008, Lu et al. 2012, Shahin 2013, Salehi and Burgueño 2018).

Numerous methods are available to classify AI algorithms (Lugar 2009) in the engineering community, however, with a general agreement to divide AI techniques into hard computing and soft computing methods (Salehi and Burgueño 2018). In hard computing, precise answers are desired using classical programming tools. Soft computing target is to find approximate answers. Most of the AI applications in civil engineering are focused on using soft computing methods to solve engineering problems and provide tools where classical mathematical methods had limited success. Soft computing algorithms mimic human reasoning methods either through physical mimicking of the human neural system such as using artificial neural networks (ANNs) (Haykin 2009) or by mimicking the approximate reasoning performed by human brains using fuzzy logic (Ross 2016). Another soft computing approach is to mimic biological systems such as swarm computing and genetic algorithms, which mimic the biological evolution, creating a set of algorithms known as evolutionary algorithms (Jang et al. 1997). The aforementioned computing methods have been successfully used in the last three decades to solve a number of nonlinear civil engineering problems.

Researchers suggested to recognize the difference between AI and data science. Data science such as big data and data mining is a branch of computer science that is focused on identifying features and recognizing patterns in data. Big data try to simplify very large data sets using algorithms. Machine learning (ML) is a subset of AI used to learn data trends and develop a model that predicts future performance. ML can be performed as supervised or unsupervised learning. Deep learning is a subset of ML that focuses on learning the features of the data.

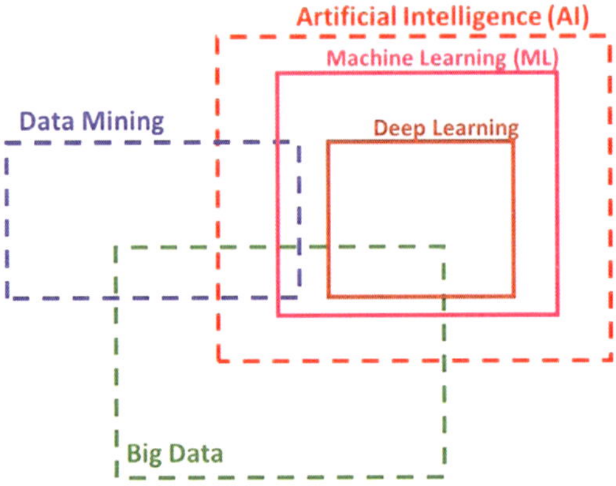

Figure 7-8. Inter-relation of the different AI techniques used in civil engineering.

Deep learning is focused on methods to enable unsupervised learning. The relationship between AI and its subsets and data science is shown schematically in Figure 7-8 after Salehi and Burgueño (2018). It is important to recognize that despite the crisp definitions of the different methods and algorithms provided previously, significant overlap takes place in applications as researchers strive to combine the capabilities of the different approaches to create intelligent algorithms and methods capable of solving field problems.

Numerous applications of AI to the field of civil infrastructure have been reported. For instance, many researchers examined the use of neural networks, fuzzy learning from examples, and neurofuzzy algorithms to extract features of vibration signals collected from bridges and high-rise building for damage detection and damage pattern recognition as part of structural health monitoring (SHM) systems (Sohn et al. 2000, Lam and Ng 2008, Mallik et al. 2016). Researchers also used evolutionary algorithms to solve construction engineering problems (Khalafallah and Abdel-Raheem 2011) and analyze soil slope stability (Ahangar-Asr et al. 2010). Furthermore, researchers examined the use of ML to design concrete mixes (Ziolkowski and Niedostatkiewicz 2019) and predicted the behavior of composite structures (Bhoopal et al. 2012) and long-term creep behavior of concrete and masonry structures (Hodhod et al. 2017, Reda Taha et al. 2004). A more comprehensive list of examples for using AI and ML methods in civil infrastructure applications can be found elsewhere (Salehi and Burgueño 2018).

The evolving research and industrial implementation of AI and ML in the last two decades, the availability of data mining and big data methods, the notable growth of computational capacities of personal computers and main stations, the significant increase in communication bandwidth, and the quantum leap in sensing technologies are enabling the civil engineering community today to make a fundamental and transformative change in future infrastructure. Integrating these tools allows the infrastructure industry to tackle the complex problem of infrastructure resilience, where a significantly high volume of data needs to be collected, communicated, and analyzed to recognize patterns in infrastructure behavior. Learned patterns then help to develop robust designs capable of overcoming future disruptive events. The integration of AI, ML, and data science with powerful computational capacities provides municipalities with high resourcefulness for in-service infrastructure that warrants rapid recovery and thus improved resilience. The integration of AI and ML technologies in smart materials creates a new generation of materials not only able to sense its environment and respond to external stressors but also able to process information and selectively interact to reduce the impact of disruption. Finally, the inference of human–infrastructure interaction capacitates resilient communities.

7.3.3 Three-Dimensional Printing

Concrete's relatively low price, as well as its ability to be mechanically engineered and poured into customizable shapes, makes it suitable for almost any structural application. Despite all these advantages, labor costs associated with this material can be extremely high, so a new way to manufacture it is a must to remain competitive. In this context, additive manufacturing of concrete (AMoC) 3D printing technology was born. Layering material on different slices makes it possible to engineer its properties in terms of composition, quantity parametrically, or printed infill pattern to meet structural requirements in a customizable and optimal way (Bos et al. 2016). There are many AMoC techniques, such as the contour crafting (Khoshnevis and Dutton 1998) AM technique, which is the most widely used for cement-based materials. Since 2013, the amount of research and construction in 3D printing projects for architectural applications has grown exponentially. This fast pace makes the field very dynamic and prone to changes. An illustration of the cumulative growth of 3D printed concrete projects in the last decade is shown in Figure 7-9.

Extensive research has been carried out in 3D technology, especially in the area of engineering, and improving 3D printable concrete to satisfy high-performance needs. In 2016, an ultrahigh-performance 3D printable concrete was developed (Gosselin et al. 2016). Research on 3D printing concrete with high-performance steel cable reinforcement was also conducted in 2017, showing an increase in ultimate deformations and strength after cracking (Bos et al. 2018). 3D printed fiber-reinforced Portland cement paste has also been printed in the past year, showing

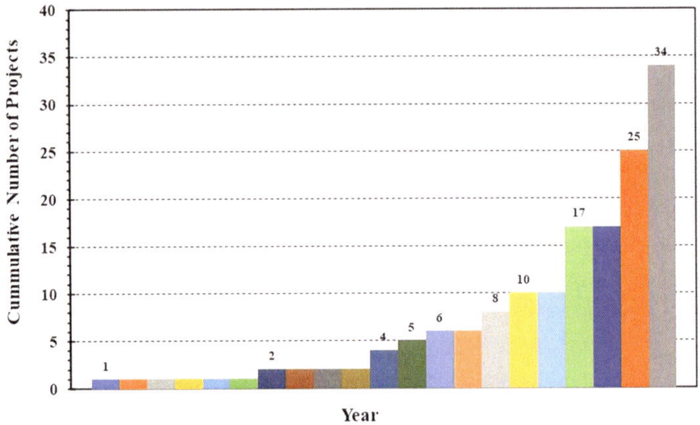

Figure 7-9. Exponential increase of large-scale additive manufacturing 3-D printing projects for construction applications since 1997.
Source: Data are extracted from Buswell et al. (2018).

compressive strength up to 100 MPa (Hambach and Volkmer 2017). Fresh properties of a novel 3D printing concrete ink using nano clay and silica fume to modify the rheology were tested, obtaining significant improvements in buildability, as well as green strength and thixotropy (Zhang et al. 2018).

Even though many large-scale structures have been printed, there are still many challenges to overcome in this emerging technology. When printing high-performance concrete, rheological and stiffening properties can limit its printability properties (Bos et al. 2016). Achieving a balance between buildability, flowability in the fresh state, and thixotropy is still a challenge for most of the mixes. Further and deeper study on the rheological properties of the mixes needs to be carried out to engineer optimum materials. Material choice, control of fresh and hardened state properties, and the inclusion of reinforcement are major challenges in current research. An example of 3D concrete printing is shown in Figure 7-10.

There are several worth-mentioning advantages of 3D printing concrete techniques compared with the traditional concrete techniques. First, there is no need for molds, which can noticeably increase the productivity and reduce the labor and indirect costs in the precast industry. In addition, the technology itself permits the material to be engineered for different needs. Finally, the possibility to print optimum topologies for each need can lead to material savings and a lower environmental footprint.

All these advantages create many potential applications in civil engineering and infrastructure fields, such as its implementation in the

Figure 7-10. 3D concrete printer.
Source: Courtesy of Mahmoud Taha.

precast industry (enabling a higher and cheaper production of some parts). An example of a 3D printed box-girder bridge segment is shown in Figure 7-11. Another application is rehabilitation of existing civil engineering structures where the access of the workforce can be a limiting factor (bridges with flooded piers). Furthermore, it can be used for nonstandard geometries in civil and residential structures that often require high planning and considerable labor costs (Bos et al. 2016). The 3D printing technique plays an important role in all those fields where traditional construction methods have been proven to be highly inefficient. Moreover, with such unique capabilities in manufacturing, 3D printing technology could be the only practical way to produce and cast all the previously discussed smart materials and to construct resilient infrastructure in the future.

Figure 7-11. 3D printed segment of the box-girder bridge.
Source: Courtesy of Mahmoud Taha.

7.4 ADVANCED SENSING TECHNOLOGY

Advanced sensing technologies play a key role in constructing new resilient infrastructure and improving the resiliency of existing infrastructure. Sensing technologies can help provide useful data to assess the real-time structural conditions of infrastructure and hence contribute to repair decisions and strategies as needed. In addition, sensing technologies could improve the resourcefulness of infrastructure resilience prior to exposure to multihazard events. Fiber-optic systems and embedded low-power wireless sensor networks are good candidates for reliable condition evaluation of local damage so that the residual safety margin of the structures can be checked after an extreme event. Satellite images, high-resolution light detection and ranging (LiDAR) radar, and computer vision images are also effective in assessing the external damages. Robotic technologies are useful for data capture of difficult-to-access sites. Other potential candidates include mobile and wearable sensors. However, there is a need to design such systems to be used in conjunction with other in-placed sensor systems during emergency events to ensure a wide coverage of sensing.

7.4.1 Fiber-Optic Sensors

By using the light scattering feature of the optic fiber, different types of fiber-optic sensors have been invented in the past three decades to quantitatively measure information surrounding the fiber such as strain, temperature, vibration, and humidity. Fiber-optic sensors have developed in two directions: point sensor and distributed sensor. The current state-of-the-art provides three types of optical fiber for SHM: (1) local fiber-optic sensors (e.g., interferometric fiber-optic sensors), (2) quasi-distributed sensors [e.g., fiber Bragg grating (FBG) sensors], and (3) distributed fiber-optic sensors (DFOSs) (Soga and Luo 2018). Interferometric fiber-optic sensors are point sensors that can detect a section that has been preprocessed on a particular location on the fiber. FBG sensors require the same predesign and premanufacturing for gratings on the fiber, but it can handle more sections on the same fiber at one channel. DFOSs can detect multiple points on the fiber with single and standard optical fiber cable.

The FBG sensor is a well-established fiber-optic-based point sensor. At present, the FBG sensor is a popular sensor because of its broad application and convenience in real cases compared with other sensors. FBGs are fabricated by photolithography by a grating mask-covered UV light [Figure 7-12(a)]. When the temperature or strain changes in the fiber, the length of the grating changes in proportion to the temperature and strain, leading to the change of the wavelength of light that is scattered; by detecting the change of scattered light, the temperature and strain information at that point can be detected. The multiple points on a single fiber can be achieved by distributing different wavelength ranges at different points. By detecting the particular wavelength range, the location of each point can be detected. In the commercialized product, normally less than 1 $\mu\varepsilon$ with 2,500 to 5,000 $\mu\varepsilon$ strain range can be achieved. However, Bragg gratings must be printed on the fiber, and each point of grating needs to be carefully attached to the place requiring detection. Because it uses multiplexing, the number of points is constrained by the limited wavelength range of each point because of the limited bandwidth of light that can be propagating on the single-mode fiber. It is typically about 10 to 20 points on each channel.

The DFOS sensor is one of the emerging technologies for monitoring structures. When a light wave travels through an optical fiber, it interacts with the constituent atoms and molecules, and the light is forced to deviate from a straight trajectory because of its nonuniformity. This deviation creates backscattering that brings a very small portion of the beam to go back to the source [Figure 7-12(b)]. When a fiber experiences a strain or temperature change, there is density fluctuation, which in turn changes the characteristics of the backscattered beam. DFOS technologies use these changes in the recovered backscattered spectrum to quantify strain, temperature, or vibration occurring along the standard optical fiber cable. By attaching an optical fiber cable to a structure or embedding it inside a structure or soil, it is possible to monitor the changes of ambient parameters of the monitored object. Typical resolutions are 10 to 30 $\mu\varepsilon$ for strain and 0.1°C to 1°C for temperature.

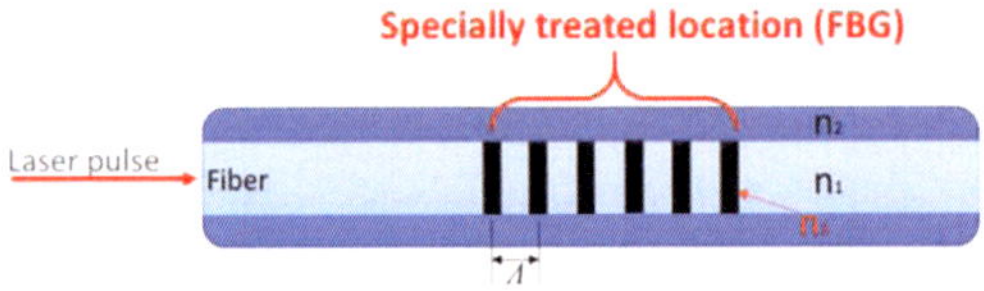

(a)

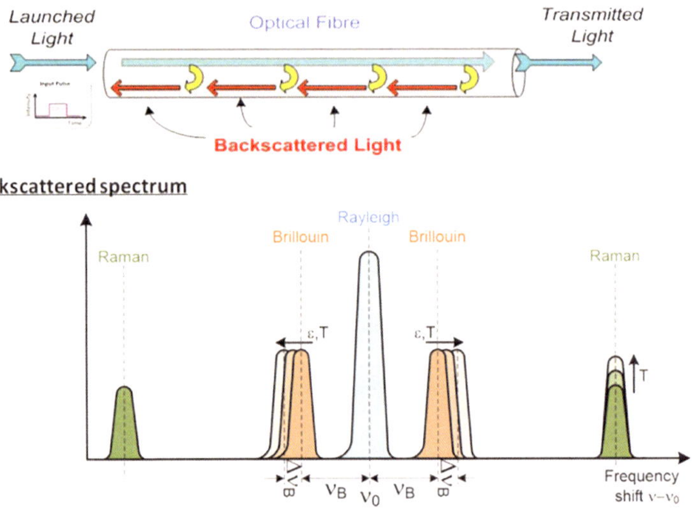

(b)

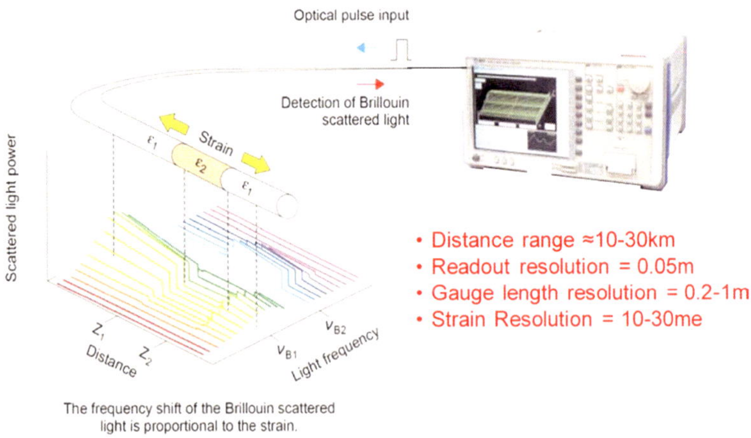

(c)

Figure 7-12. (a) Fiber Bragg grating, (b) backscattered lights from different locations of the fiber, and (c) fiber with the DFOS analyzer.

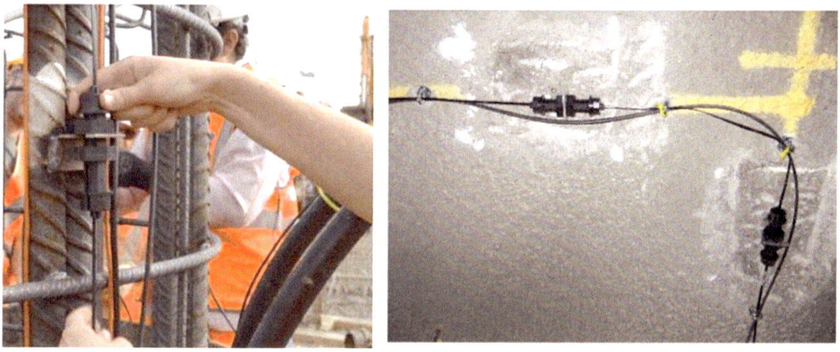

Figure 7-13. Installation of optical fiber cables for monitoring distributed strains in the pile and tunnel.

The main advantage of DFOS is its high sensitivity over large distances (a few to tens of kilometers) and the ability to interface with a wide range of measurements in a distributed manner (data samples of every 2 cm to 1 m) [Figure 7-11(c)]. The system provides thousands of strain gauges, thermocouples, or accelerometers along a single, standard fiber-optic cable, which results in a low-cost monitoring solution considering the density of the data obtained. Moreover, the deterioration rate of the glass optical fiber material is slow. Hence, it is considered to be an ideal measurement method for long-term monitoring by embedding it into structures (Figure 7-13). Further examples can be found in Kechavarzi et al. (2016).

7.4.2 Digital Image Sensing

Visual inspection is commonly used for detecting anomalies such as cracks, spalling, and staining. Inspectors visually compare the images to determine whether any anomalies have arisen between inspections. The goal of change detection using recent computer vision technologies is to identify the regions of changes between multiple images of the same scene taken at different times. Because image collections often become large and difficult to organize and browse, improving the ways to access old images can lead to substantial progress in the effectiveness of monitoring. An automatic tool that combines a large number of images into a single high-quality wide-angle composite view is useful to organize a large collection of images for inspection.

Structure from motion (SFM) is a system that simultaneously recovers a 3D point cloud model and camera positions using only 2D images (Lowe 1999). A 3D model construction algorithm typically starts with an algorithm to extract features and their associated descriptor vectors for each image.

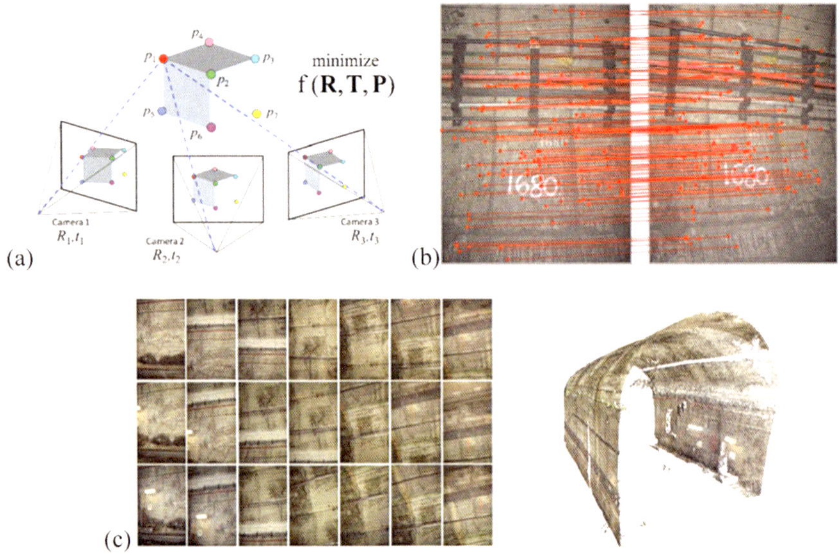

Figure 7-14. (a) SFM concept to create a 3D model, (b) scale-invariant feature transform matching algorithm to identify the same points in two images, and (c) 3D construction of a Barcelona Metro tunnel.

The object appearing in the image can then be constructed as a three-dimensional one using a surveying technique (Figure 7-14). Some SFM systems have recently been commercialized. The software enables users to create a 3D point cloud model from uploaded photographs and allows users to browse and navigate through the photographs. It can simultaneously cope with the free camera motion and more complex geometry of the scene. The accuracy is becoming close to that of a traditional laser scanning method. Images are registered by transforming image sets from varying and unknown coordinate systems into one single coordinate system. By assigning spatial coordinates to each image, it is possible to identify regions of change in images of the same scene taken at different times. Such systems can reduce the workload for visual inspection significantly versus competing techniques, facilitating high frequency, effective inspections (Chaiyasarn et al. 2016, Stent et al. 2016 for tunnel inspection). Further work is needed to capture material information when geometric data are acquired. Ideally, materials should be identified nondestructively, and there is a need to create a database of materials so that semantics of a captured scene can be harvested via various computer learning techniques (e.g., deep learning).

Digital Image Correlation (DIC) consists of taking a series of photographs in sequence and calculating the velocity field by comparing them. It has been implemented in the field for SHM of infrastructure assets and a precision of 0.008 mm has been reported for field application using a standard digital camera (Alhaddad et al. 2019). Challenges in the field become more rigorous than in the laboratory when considering the real-world illumination conditions. Lighting is nonhomogeneous because of the varying usage of flashlights, and shadows and darkness are observed in local areas. The most common application of DIC has been on bridge monitoring with deployment over a short period and for applications that need high levels of frequency of images, such as vibration monitoring.

7.4.3 LiDAR

Light detection and ranging (LiDAR) radars provide for a remote sensing technique that measures distance with pulsed laser light and collects the measurement points to build a 3D point cloud model of the scanned object, as shown in Figure 7-15. Terrestrial laser scanners are geomatic devices equipped with a laser beam and precise servomotors. They collect 3D point data and can carry out similar surveying operations to total stations. In recent years, there have been attempts to carry out LiDAR measurements from unmanned aerial vehicles (UAVs).

Comparing different point clouds taken at different times is challenging (Acikgoz et al. 2017). In particular, registering clouds to the same coordinate system is critical, because slight errors in registration may render further comparisons unreliable. Currently, the existing approaches to comparing point clouds can be grouped under two approaches. The first approach computes distances between clouds. Existing methods include the determination of the closest point-to-point distance between two-point clouds. The second approach is the computation of displacement vectors by identifying corresponding sections of two-point clouds. These methods determine the rigid body rototranslation matrix of the displaced object between two scans. In this method, the distances between the corresponding object in two clouds are minimized with an optimization algorithm, starting with an initial guess based on the closest distance between clouds.

Figure 7-15. LiDAR device and point cloud of the masonry arch structure.

7.4.4 Wireless Sensor Network

For monitoring large-scale infrastructure, introducing a wired monitoring system takes a great deal of time and cost. In contrast, wireless sensor network (WSN) is a very attractive technology because cable installation is not necessary. Combining it with high-precision and low-cost sensors enables large-scale monitoring, which was not possible in the past. The advantages of WSN to monitor the behavior of civil infrastructure are as follows: (1) multiple numbers and types of sensors can be deployed without installing cables, (2) sensing in areas difficult to connect possible, (3) because it is possible to obtain a wide range of data in real time, decision making can be made quickly, and (4) equipped with an internal battery, each node is self-sustainable over long periods of time without battery change. An advanced WSN system contains mesh network capability, and hence, any node can behave as a sensor node and a relay node. This gives the maximum flexibility to nonradio professionals to deploy the system with minimal effort (time and cost).

The performance of WSN hardware and software has dramatically improved in recent years owing to innovative sensors (Figure 7-16), not only in data collection and abnormal signal detection but also in data compression processing, data analysis, and network processing. Data communication is optimized, and power consumption is suppressed. With low-power high-performance central-processing units, it is now possible to perform data processing at the sensing level, which in turn saves energy by sending only the necessary data. WSNs have already been recognized as a low-cost and resource-efficient alternative to these wired technologies for a number of deployments in structures, enabling numerous monitoring opportunities that would hardly even be thought of a decade ago.

7.4.5 Satellite Images

Satellites take images of the earth. The WorldView-3 satellite captures (1) panchromatic data of 450 to 800 nm wavelengths at 0.30 m resolution;

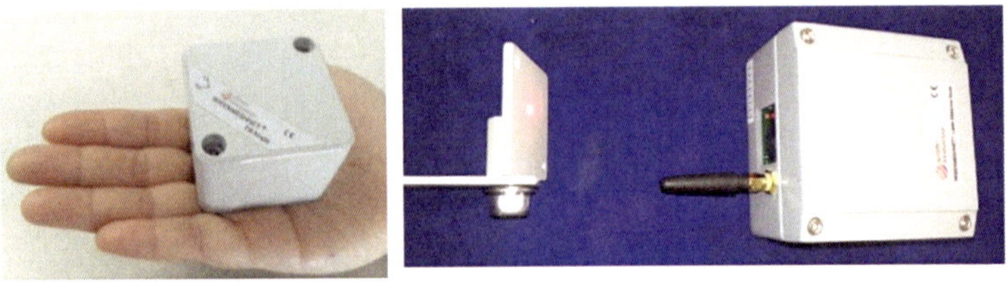

(a) (b)

Figure 7-16. WSN sensor examples: (a) tilt mote (Wisen innovation), and (b) laser distance mote (Wisen innovation).

(2) multispectral data (spectral bands of 50 to 60 nm wide between 400 and 1,000 nm wavelengths at 1.25 m resolution; and (3) shortwave infrared data of 1,000 to 2,500 nm wavelengths at 4 m resolution. Passive sensors are mounted on the satellite to capture the solar radiation reflected from the earth surface. This is in contrast to an active sensor capture system, which sends an energy wave to the earth surface and records the reflected wave. Using techniques similar to DIC, displacement fields can be estimated to be less than 20 m intervals. Horizontal displacements as small as 0.2 to 0.3 m can be measured.

Synthetic aperture radar (SAR) can produce high-resolution radar images of the earth surface and can typically cover a surface area of between 2,500 and 10,000 km^2, by mounting the system on a satellite or UAV. Because SAR uses the microwave band in the broad radio spectrum, it has a day-and-night imaging capability and the ability to penetrate through rain/water clouds. The potential information in the phase of SAR complex images has led to a technology called interferometric SAR (InSAR). Satellite-based multitemporal interferometric synthetic aperture radar (MT-InSAR) is a monitoring technique capable of extracting cumulative surface displacement measurements with millimeter accuracy (Figure 7-17).

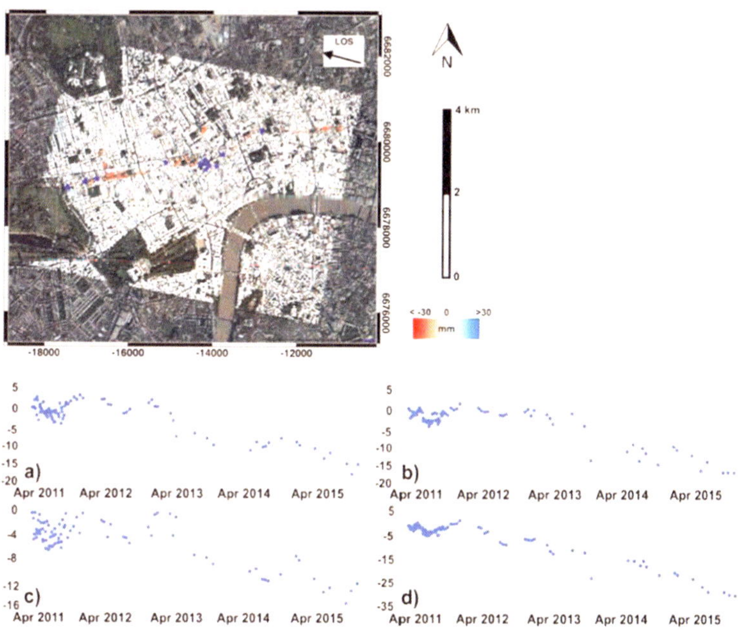

Figure 7-17. Cumulative displacement map along the Crossrail tunnel construction alignment in London from April 2011 to December 2015. The negative value is subsidence. Settlement–time curves at four locations along the alignment.
Source: Milillo et al. (2018).

7.4.6 Augmented Reality

Augmented reality (AR) refers to an enriched real world with a complimenting virtual world. In the case of virtual reality (VR), the real world is replaced by virtual objects and systems. In contrast, AR enhances the real world by anchoring virtual information into it. The beginning of AR dates back to the 1960s, where head-mounted displays were used to present 3D graphics (Sutherland 1968). Such early inventions had specific problems with resolution, brightness, and field of view (Azuma et al. 2001). A 1997 survey on AR provided the field's uses, challenges, and developments (Azuma 1997). Since then, the AR ecosystem has provided applications in various fields, including construction, structural inspection, evaluation, and renovation. In 1996, an AR system was developed that showed the location of columns behind a finished wall, the location of rebars inside one of the columns, and performed a structural analysis of the column (Webster et al. 1996). In 1999, a similar AR testbed system was used to address spaceframe construction (Feiner et al. 1999). These systems focused on demonstrating the potential of AR's X-ray vision with a lab-based approach. In addition, challenges related to occlusion were experienced with early age devices. Occlusion, otherwise known as obstruction of real objects by virtual ones, was not possible in conventional optical see-through displays.

Recent achievements have shown promise for using AR in civil engineering/infrastructure. For example, an AR camera (ARCam) has been successfully used to inspect the column anchor bolt positions before installing a steel column and to assess its plumpness after installation (Shin and Dunston 2009). The ARCam demonstrated time-related advantages when compared with the conventional total station, thus increasing productivity and reducing cost. In addition, measurement precision was found in compliance with standard tolerances. Further advancements were subsequently made in the AR industry introducing devices such as the Google glass, Moverio smart glasses, and the current Microsoft HoloLens (Cass and Choi 2015, Moreu et al. 2017) [Figures 7-18(a and b)]. Recent advancements in AR have been successfully advancing the area of interaction between humans and the built environment, especially following disasters, which translates into increased resilience after disasters. Figure 7-17(b) shows the 3D data collected data.

7.4.7 Unmanned Aerial Systems

Unmanned aerial systems (UASs), also known as UAVs or drones, refer to flying robots or mechanisms. UASs are used by civil engineers

(a) (b)

Figure 7-18. Augmented reality: (a) HoloLens from Microsoft, and (b) HoloLens data from 3D meshing of the structural element.

to inspect and access large infrastructure, which previously was not possible because of their elevated cost, which has now become reasonable for research and development. In recent years, UAS platforms have become more popularly used in several areas, including inspection, photography, surveillance, data collection, and remote sensing (Blanks 2016). Using the UAS offers more flexibility and access to the structures, which was previously not possible (Cummings et al. 2017). Owing to their agility, UASs have found their use in several applications such as disaster management (Restas 2015), oil spill surveillance and detection (Allen and Walsh 2008), soil erosion monitoring (d'Oleire-Oltmanns et al. 2012), forest ecosystem and biodiversity monitoring (Getzin et al. 2014), and deforestation detection (Paneque-Gálvez et al. 2014). Researchers have used UAS-based systems for bridge monitoring in recent years. For example, structural engineers and managers are interested in UAS-based systems implementing image processing and 3D mapping of the structures being inspected (Yoon et al. 2018). UASs have also been used in SHM in the form of novel methods like checking the quality of the concrete by tap testing it with a hammer mounted to a UAS because of its accessibility (Mason et al. 2016, Moreu et al. 2018). More recently, engineers have explored adding sensing devices such as lasers to collect dynamic responses of structures during dynamic events (Figure 7-19) (Garg et al. 2019). UAS has been identified as promising technology to assess the condition of infrastructure after natural or human disasters, victim searches, structural assessment after disasters, and remote sensing. It is expected that resilient structures and communities will be assisted by UAS to quantify the damage of structures following disasters.

Figure 7-19. UAS used for SHM and infrastructure monitoring.

7.5 FIELD IMPLEMENTATION OF EMERGING TECHNOLOGIES

The previous sections provided a brief description of a few emerging materials and technologies that have great potential for improving the resilience of existing and future infrastructure. To illustrate the benefits of such materials and technologies, we consider an example of a bridge structure where shape memory alloy (SMA) seismic dampers are installed and where a wireless distributed sensor network is deployed. Furthermore, damage to the bridge structure is detected and quantified using the distributed sensor network empowered with AI algorithms (Reda Taha and Lucero 2005). Finally, self-healing materials (e.g., concrete and composite elements) are included as part of the bridge elements. The bridge response to a disruptive earthquake event will showcase the contribution of the emerging materials and technologies to the bridge resilience. The sensor network will enable continuous assessment of the bridge performance over time and thus provide the necessary knowledge to offer resources for bridge maintenance. This *resourcefulness* will allow the bridge to retain an appropriate level of functionality before the disruptive event. When the earthquake impacts the bridge, the SMA seismic dampers will reduce the bridge functionality loss, providing necessary *redundancy* to the bridge performance. The sensor network supported by the AI algorithm will offer additional *resourcefulness* and provides *rapidity* in realizing the bridge's health and identify the specific needs to repair. Finally, the inclusion of self-healing materials will improve the *rapid* recovery of the bridge functionality. It is evident from the aforementioned example that incorporating emerging materials and technologies contributes to the main elements of resilience and improves the bridge resilience compared with conventional bridges where none of the emerging technologies is considered. The selection of the technology to be included in the infrastructure is a function of the infrastructure type and the specific hazard to mitigate.

Despite their large potential benefits to resilience of infrastructure, the expectations from some emerging technologies vary with time following the Gartner hype cycle, as shown in Figure 7-20. The Gartner hype cycle consists of five distinct stages with varying levels of expectations, namely, innovative trigger, peak of inflated expectation, trough of disillusionment, slope of enlightenment, and plateau of productivity (Hansford 2016). Depending on its first introduction and research/implementation efforts, the current expectation for any emerging technology can be determined. Maximum benefits for technologies in industries and to communities are typically achieved in the final stage in the curve. As noticed in the figure, many emerging technologies are still at the developmental stages. For instance, the technology of 3D printing of a whole building is still at its first stage (*innovative trigger*) with limited R&D efforts and startup companies. Technologies related to IoT are currently in the second stage (*peak of inflated expectations*), whereas fiber-optic sensors and carbon fibers are advancing in the fourth stage (*slope of enlightenment*). More efforts are definitely required to transfer many of the emerging technologies pertaining to infrastructure resilience to the productivity stage in the coming years.

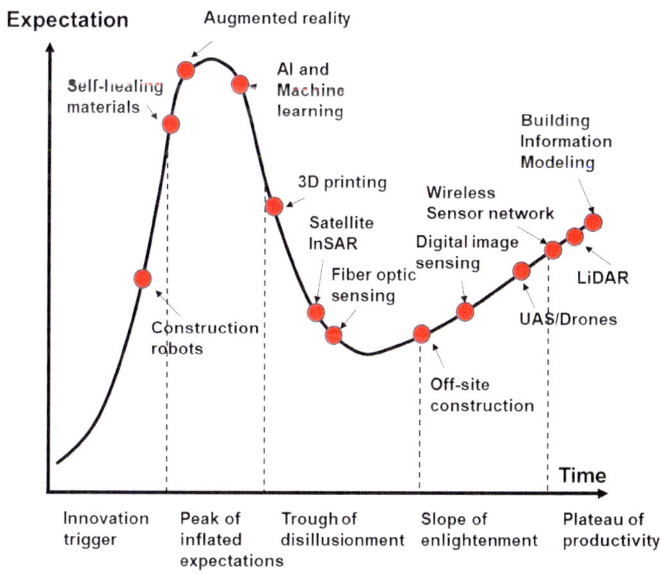

Figure 7-20. Gartner hype cycle for emerging technologies.
Source: New Civil Engineer (2016).

Because most of these technologies are still in the research phase with little to no field implementations, there is an essential need for developing a roadmap to promote field implementation of such new technologies. An example field implementation roadmap is shown in Figure 7-20. The roadmap consists of several phases that promote technology transfer from the lab scale at the research phase to the city scale with extensive field implementations. Such a roadmap advances useful technologies to the plateau of the productivity stage in the hype cycle depicted in Figure 7-21.

Each phase in the roadmap is associated with a set of inquiries that shall be addressed for each emerging technology to determine its feasibility for large-scale implementation. In the first phase, it is important to recognize the required material criteria that can lead to improved resiliency, to realize the necessary level of maturity/readiness for field application, and to develop material standards and testing specifications. In contrast, for robust sensing techniques, it is important to recognize the basic need for the sensors, identify methods to make them robust and reliable, and develop standards to enable interoperability. The second phase involves data analysis and interpretation in real time for measuring the efficiency of the emerging technologies. In this phase, methods for designing, constructing, and monitoring infrastructure with the new emerging technologies shall be developed. In addition, advanced techniques for data

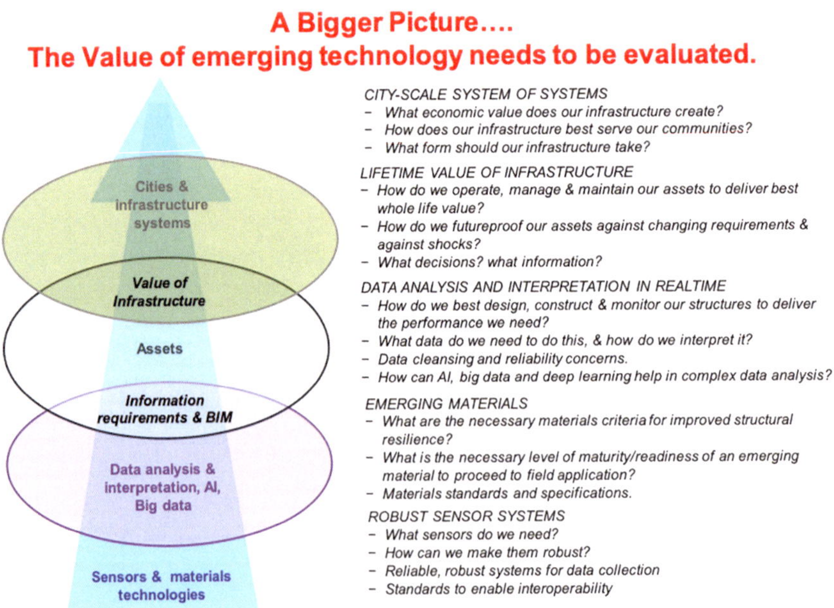

Figure 7-21. Large-scale implementation roadmap of emerging technologies for resilient infrastructure.

recording, analysis, and cleansing shall also be established. In the third phase, the lifetime value for the emerging technology and its impact on resiliency of individual infrastructure shall be recognized, including developing approaches for long-term operation, management, maintenance, decision making, and asset protection. The fourth and final phase at the city level shall address concerns related to the economic and social impact of implementing resilience emerging technologies in our communities.

REFERENCES

Abdel-Emam, M., E. Soliman, A. Nassr, W. Khair-Eldeen, and A. Abd-Elshafy. 2018. "Dynamic behavior of textile reinforced polymer concrete using split Hopkinson pressure bar." In *Proc., Int. Congr. on Polymers in Concrete*. Washington, DC, Springer, ISBN-13: 978-3319781747.

Abel, M.-L., J. F. Watts, and R. P. Digby. 2004. "The influence of process parameters on the interfacial chemistry of γ-GPS on aluminum: A review." *J. Adhes.* 80 (4): 291–312.

ACI (American Concrete Institute). 2009. *Guide for the use of polymers in concrete.* ACI 548.1R-09. Farmington Hills, MI: ACI.

Acikgoz, S., K. Soga, and J. Woodhams. 2017. "Evaluation of the response of a vaulted masonry structure to differential settlements using point cloud data and limit analyses." *Constr. Build. Mater.* 150 (30): 916–931.

Adam, B., and I. F. C. Smith. 2008. "Active tensegrity: A control framework for an adaptive civil-engineering structure." *Comput. Struct.* 86 (23–24): 2215–2223.

Adeli, H., and S. L. Hung. 1994. *Machine learning: Neural networks, genetic algorithms and fuzzy systems.* New York: Wiley.

Ahangar-Asr, A., A. Faramarzi, and A. A. Javadi. 2010. "A new approach for prediction of the stability of soil and rock slopes." *Eng. Comput.* 27 (7): 878–893.

Ahn, T.-H., and T. Kishi. 2010. "Crack self-healing behavior of cementitious composites incorporating various mineral admixtures." *J. Adv. Concr. Technol.* 8 (2): 171–186.

Aïssa, B., D. Therriault, E. Haddad, and W. Jamroz. 2012. "Self-healing materials systems: Overview of major approaches and recent developed technologies." *Adv. Mater. Sci. Eng.* 2012: 854203.

Alaneme, K. K., and E. A. Okotete. 2016. "Reconciling viability and cost-effective shape memory alloy options—A review of copper and iron based shape memory metallic systems." *Eng. Sci. Technol.* 19 (3): 1582–1592.

Aldersey-Williams, H. 2004. "Towards biomimetic architecture." *Nat. Mater.* 3 (5): 277–279.

Alhaddad, M., M. Dewhirst, K. Soga, and M. Devriendt. 2019. "A new photogrammetric system for high-precision monitoring of tunnel deformations." In *Proc., Inst. Civ. Eng. Transp.* 172 (2): 81–93.

Allen, J., and B. Walsh. 2008. "Enhanced oil spill surveillance, detection and monitoring through the applied technology of unmanned air systems." In *Proc., Int. Oil Spill Conf.*, 113–120. Washington, DC: American Petroleum Institute.

Al-Sabagh, A., E. Taha, U. F. Kandil, A. E. Awadallah, G. Nasr, and M. M. Reda Taha. 2017. "Monitoring moisture damage propagation in GFRP composites using carbon nanoparticles." *Polymers* 9: 94.

Azuma, R. T. 1997. "A survey of augmented reality." *Presence* 6 (4): 355–385.

Azuma, R., Y. Baillot, R. Behringer, S. Feiner, S. Julier, and B. MacIntyre. 2001. "Recent advances in augmented reality." *IEEE Comput. Graphics Appl.* 21 (6): 34–47.

Berry, D. H., and A. Namkanisorn. 2005. "Fracture toughness of a silane coupled polymer–metal interface: Silane concentration effects." *J. Adhes.* 81: 347–370.

Bhoopal, R. S., R. Singh, and P. K. Sharma. 2012. "Adaptative neuro-fuzzy inference system for prediction of effective thermal conductivity of polymer–matrix composites." *Model. Numer. Simul. Mater. Sci.* 2: 43–50.

Blaiszik, B. J., S. L. B. Kramer, S. C. Olugebefola, J. S. Moore, N. R. Sottos, and S. R. White. 2010. "Self-healing polymers and composites." *Annu. Rev. Mater. Res.* 40: 179–211.

Blanks, M. 2016. "UAS applications." In *Introduction to unmanned aircraft systems*, edited by D. M. Marshall, R. K. Barnhart, E. Shappee, and M. Most. Boca Raton, FL: CRC Press.

Boh, B., and B. Šumiga. 2008. "Microencapsulation technology and its applications in building construction materials (tehnologija mikrokapsuliranja in njena uporaba v gradbenih materialih)." *RMZ–Mater. Geoenviron.* 55 (3): 329–344.

Bos, F. P., Z. Y. Ahmed, R. J. M. Wolfs, and T. A. M. Salet. 2018. "3D printing concrete with reinforcement." In *high tech concrete: Where technology and engineering meet*, edited by D. A. Hordijk and M. Luković, 2484–2493. Cham, Switzerland: Springer.

Bos, F., R. Wolfs, Z. Ahmed, and T. Salet. 2016. "Additive manufacturing of concrete in construction: Potentials and challenges of 3D concrete printing." *Virtual Phys. Prototyp.* 11 (3): 209–225.

Brown, E. N., N. R. Sottos, and S. R. White. 2002. "Fracture testing of a self-healing polymer composite." *Exp. Mech.* 42 (4): 372–379.

Buswell, R. A., W. R. Leal de Silva, S. Z. Jones, and J. Dirrenberger. 2018. "3D printing using concrete extrusion: A roadmap for research." *Cem. Concr. Res.* 112: 37–49.

Cailleux, E., and V. Pollet. 2009. "Investigations on the development of self-healing properties in protective coatings for concrete and repair mortars." In Vol. 28 of *Proc., 2nd Int. Conf. on Self-Healing Materials*. Chicago, IL, USA, 28 June–1 July 2009.

Caruso, M. M., B. J. Blaiszik, S. R. White, N. R. Sottos, and J. S. Moore. 2008. "Full recovery of fracture toughness using a nontoxic solvent-based self-healing system." *Adv. Funct. Mater.* 18 (13): 1898–1904.

Caruso, M. M., D. A. Delafuente, V. Ho, N. R. Sottos, J. S. Moore, and S. R. White. 2007. "Solvent-promoted self-healing epoxy materials." *Macromolecules* 40 (25): 8830–8832.

Cass, S., Choi, C.Q., 2015. "Google glass, hololens, and the real future of augmented reality." *IEEE Spectr.* 52 (3): 18. Accessed March 12, 2021. http://spectrum.ieee.org/consumer-electronics/audiovideo/google-glass-hololens-and-the-real-future-of-augmented-reality.

Celestine, A. N., N. R. Sottos, and S. R. White. 2015. "Autonomic healing of PMMA via microencapsulated solvent." *Polymer* 69: 241–248.

Chaiyasarn, K., T. K. Kim, F. Viola, R. Cipolla, and K. Soga. 2016. "Distortion-free image mosaicing for tunnel inspection based on robust cylindrical surface estimation through structure from motion." *J. Comput. Civil Eng.* 30 (3): 04015045.

Choi, E., Y.-S. Chung, D.-H. Choi, and R. DesRoches. 2012. "Seismic protection of lap-spliced RC columns using SMA wire jackets." *Mag. Concr. Res.* 64 (3): 239–252.

Cummings, A. R., A. McKee, K. Kulkarni, and N. Markandey. 2017. "The rise of UAVs." *Photogramm. Eng. Remote Sens.* 83 (4): 317–325.

Daghash, S. M., and O. E. Ozbulut. 2016. "Characterization of superelastic shape memory alloy fiber-reinforced polymer composites under tensile cyclic loading." *Mater. Des.* 111: 504–512.

Daghash, S. M., and O. E. Ozbulut. 2018. "Tensile and fatigue behavior of polymer composites reinforced with superelastic SMA strands." *Smart Mater. Struct.* 27 (6): 065003.

Dalal, M., O. Goumairi, and A. El Maliki. 2017. "Study of the internal confinement of concrete reinforced (in civil engineering) with woven reinforcement." In *Proc., 17th World Textile Conf. AUTEX 2017—Textiles—Shaping the Future*. 29–31 May 2017, Corfu, Greece.

d'Oleire-Oltmanns, S., I. Marzolff, K. D. Peter, and J. B. Ries. 2012. "Unmanned aerial vehicle (UAV) for monitoring soil erosion in Morocco." *Remote Sens.* 4 (11): 3390–3416.

Dry, C. 1994. "Matrix cracking repair and filling using active and passive modes for smart timed release of chemicals from fibers into cement matrices." *Smart Mater. Struct.* 3 (2): 118–123.

Du, Y., X. Zhang, L. Liu, F. Zhou, D. Zhu, and W. Pan. 2018. "Flexural behaviour of carbon textile-reinforced concrete with prestress and steel fibres." *Polymers* 10 (1): 98.

Duncanson, W. J., T. Lin, A. R. Abate, S. Seiffert, R. K. Shah, and D. A. Weitz. 2012. "Microfluidic synthesis of advanced microparticles for encapsulation and controlled release." *Lab Chip* 12 (12): 2135–2145.

Eastman, C. 1975. "The use of computers instead of drawings in building design." *AIA J.* 63 (3): 46–50.

Eastman, C., P. Teicholz, R. Sacks, and K. Liston. 2012. "A guide to building information modeling for owners, managers, architects, engineers and contractors." *Australas. J. Constr. Econ. Build.* 12 (3): 101–102.

Emiroglu, M., A. Douba, R. A. Tarefder, U. F. Kandil, and M. R. Taha. 2017. "New polymer concrete with superior ductility and fracture toughness using alumina nanoparticles." *J. Mater. Civ. Eng.* 29 (8): 04017069.

Feiner, S., A. Webster, B. MacIntyre, and T. Hollerer. 1999. "Augmented reality for construction." Accessed January 7, 2020. http://www1.cs.columbia.edu/graphics/projects/arc/arc.html.

Feng, N., L. Liu, Y. Liu, and J. Leng. 2015. "Characteristics of multi-functional composites using elastomer embedded with shape memory alloy wires." *Mater. Des.* 88: 75–81.

Garg, P. R., F. Moreu, A. I. Ozdagli, M. M. Reda Taha, and D. L. Mascarenas. 2019. "Non-contact dynamic displacement measurement of structures using a moving laser Doppler vibrometer." *J. Bridge Civ. Eng.* Accessed March 22, 2021. https://ascelibrary.org/doi/abs/10.1061/%28ASCE%29BE.1943-5592.0001472.

Getzin, S., R. S. Nuske, and K. Wiegand. 2014. "Using unmanned aerial vehicles (UAV) to quantify spatial gap patterns in forests." *Remote Sens.* 6 (8): 6988–7004.

Ghafari, E., Y. Yuan, C. Wu, T. Nantung, and N. Lu. 2018. "Evaluation the compressive strength of the cement paste blended with supplementary cementitious materials using a piezoelectric-based sensor." *Constr. Build. Mater.* 171: 504–510.

Gladman, A. S., A.-D. N. Celestine, N. R. Sottos, and S. R. White. 2015. "Autonomic healing of acrylic bone cement." *Adv. Healthcare Mater.* 4: 202–207.

Gosselin, C., R. Duballet, P. Roux, N. Gaudillière, J. Dirrenberger, and P. Morel. 2016. "Large-scale 3D printing of ultra-high performance concrete—A new processing route for architects and builders." *Mater. Des.* 100: 102–109.

Gries, T., M. Raina, T. Quadflieg, and O. Stolyarov. 2016. "Manufacturing of textiles for civil engineering applications." In *Textile fiber composites in civil engineering*, edited by T. Triantafillou, 3–24. Cambridge, UK: Woodhead Publishing.

Hambach, M., and D. Volkmer. 2017. "Properties of 3D-printed fiber-reinforced Portland cement paste." *Cem. Concr. Compos.* 79: 62–70.

Hansford, M. 2016. "New civil engineer." Accessed January 7, 2020. https://www.newcivilengineer.com/delivering-differently-emerging-technologies/10001703.article.

Haykin, S. 2009. *Neural networks and learning machines*. 3rd ed. New York: Pearson.

Hodhod, O., T. E. Said, and A. M. Ataya. 2017. "Prediction of creep in concrete using genetic programming hybridized with ANN." *Comput. Concr.* 21 (5): 513–523.

Hung, C. C., and W. M. Yen. 2014. "Experimental evaluation of ductile fiber reinforced cement-based composite beams incorporating shape memory alloy bars." *Procedia Eng.* 79: 506–512.

Jamshaid, H., R. Mishra, J. Militký, and M. T. Noman. 2018. "Interfacial performance and durability of textile reinforced concrete." *J. Text. Inst.* 109 (7): 879–890.

Jang, J. S. R., C. T. Sum, and E. Mizutani. 1997. *Neuro-fuzzy and soft computing: A computational approach to learning and machine intelligence*. New York: Pearson.

Jani, J. M., M. Leary, A. Subic, and M. A. Gibson. 2014. "A review of shape memory alloy research, applications and opportunities." *Mater. Des.* 56: 1078–1113.

Jin, H., G. A. Miller, N. R. Sottos, and S. R. White. 2011. "Fracture and fatigue response of a self-healing epoxy adhesive." *Polymer* 52 (7): 1628–1634.

Jonkers, H. M., and E. Schlangen. 2008. "Development of a bacteria-based self-healing concrete." In *Proc., Int. FIB Symp.*, Amsterdam, Netherlands, 425–430.

Jonkers, H. M., A. Thijssen, G. Muyzer, O. Copuroglu, and E. Schlangen. 2010. "Application of bacteria as self-healing agent for the development of sustainable concrete." *Ecol. Eng.* 36 (2): 230–235.

Kechavarzi, C., K. Soga, N. de Battista, L. Pelecanos, M. Elshafie, and R. J. Mair. 2016. *Distributed optical fiber sensing for monitoring geotechnical infrastructure—A practical guide*. London: ICE Publishing.

Kensek, K., and D. Noble. 2014. *Building information modeling: BIM in current and future practice*. Hoboken, NJ: Wiley.

Khalafallah, A., and M. Abdel-Raheem. 2011. "Electimize: New evolutionary algorithm for optimization with application in construction engineering." *J. Comput. Civil Eng.* 25 (3): 192–201.

Khoshnevis, B., and R. Dutton. 1998. "Innovative rapid prototyping process makes large sized, smooth surfaced complex shapes in a wide variety of materials." *Mater. Technol.* 13 (2): 53–56.

Knippers, J., and T. Speck. 2012. "Design and construction principles in nature and architecture." *Bioinspir. Biomim.* 7 (1): 015002.

Koutas, L. N., Z. Tetta, D. A. Bournas, and T. C. Triantafillou. 2019. "Strengthening of concrete structures with textile reinforced mortars: State-of-the-art review." *J. Compos. Constr.* 23 (1): 03118001.

Kuang, Y., and J. Ou. 2008. "Self-repairing performance of concrete beams strengthened using superelastic SMA wires in combination with adhesives released from hollow fibers." *Smart Mater. Struct.* 17 (2): 025020.

Kulas, C. 2015. "Actual applications and potential of textile-reinforced concrete." In *Proc., 17th Int. Cong. of the GRCA*. Radison Blu, Dubai.

Lam, H. F., and C. T. Ng. 2008. "The selection of pattern features for structural damage detection using an extended Bayesian ANN algorithm." *Eng. Struct.* 30: 2762–2770.

Lepenies, I. G., M. Richter, and B. W. Zastrau. 2008. "A multi-scale analysis of textile reinforced concrete structures." In *Proc., Applied Mathematics and Mechanics*. https://doi.org/10.1002/pamm.200810553 Accessed April 8, 2021.

Lin, C. B., S. Lee, and K. S. Liu. 1990. "Methanol-induced crack healing in poly(methyl methacrylate)." *Polym. Eng. Sci.* 30: 1399–1406.

Lowe, D. G. 1999. "Object recognition from local scale-invariant features." In *Proc., 7th IEEE Int. Conf. on Computer Vision*, 1150–1157, Kerkyra, Greece, IEEE.

Lu, P., S. Chen, and Y. Zheng. 2012. "Artificial intelligence in civil engineering." *Math. Probl. Eng.* 2012: 145974.

Lugar, G. F. 2009. *Artificial intelligence: Structures and strategies for complex problem solving*. 6th ed. New York: Pearson.

Mallik, N., A. Wali, and N. Kuri. 2016. "Damage location identification through neural network learning from optical fiber signal for structural health monitoring." In *Proc., 5th Int. Conf. on Mechatronics and Control Engineering*, 157–161. New York: ACM.

Mason, J. D., E. T. Ayorinde, D. D. Mascarenas, and F. Moreu. 2016. *Tap testing hammer using unmanned aerial systems (UASs)*. No. LA-UR-16-26204. Los Alamos, NM: Los Alamos National Laboratory.

Mihashi, H., Y. Kaneko, T. Nishiwaki, and K. Otsuka. 2001. "Fundamental study on development of intelligent concrete characterized by self-healing capability for strength." *Trans. Jpn. Concr. Inst.* 22: 441–450.

Milillo, P., G. Giardina, M. J. DeJong, D. Perissin, and G. Milillo. 2018. "Multi-temporal InSAR structural damage assessment: The London Crossrail case study." *Remote Sens.* 10 (2): 287.

Moreu, F., E. Ayorinde, J. Mason, C. Farrar, and D. Mascarenas. 2018. "Remote railroad bridge structural tap testing using aerial robots." *Int. J. Intell. Rob. Appl.* 2 (7–8): 67–80.

Moreu, F., B. Bleck, S. Vemuganti, D. Rogers, and D. Mascarenas. 2017. "Augmented reality tools for enhanced structural inspection." In *Proc., 11th Int. Workshop on Structural Health Monitoring*. Stanford, California. https://web.stanford.edu/group/sacl/workshop/IWSHM2017/naccommodation2.html

Mosallam, A. 2011. *Design guide for FRP composite connection*. Reston, VA: ASCE.

Munch, E., M. E. Launey, D. H. Alsem, E. Saiz, A. P. Tomsia, and R. O. Ritchie. 2008. "Tough, bio-inspired hybrid materials." *Science* 322 (5907): 1516–1520.

Naboni, R., and I. Paoletti. 2015. *Advanced customization in architectural design and construction*. Cham, Switzerland: Springer.

Namkanisorn, A., A. Ghatak, M. K. Chaudhury, and D. H. Berry. 2001. "A kinetic approach to study the hydrolytic stability of polymer-metal adhesion." *J. Adhes. Sci. Technol.* 15: 1725–1745.

Orlowsky, J., and M. Raupach. 2011. "Textile reinforced concrete—From research to application." *Cement, Wapno, Beton* 16 (6): 323–331.

Ozbulut, O. E., S. Daghash, and M. M. Sherif. 2016. "Shape memory alloy cables for structural applications." *J. Mater. Civ. Eng.* 28 (4): 04015176.

Ozbulut, O. E., and S. Hurlebaus. 2012. "A comparative study on the seismic performance of superelastic-friction base isolators against near-field earthquakes." *Earthquake Spectra* 28 (3): 1147–1163.

Paneque-Gálvez, J., M. K. McCall, B. M. Napoletano, S. A. Wich, and L. P. Koh. 2014. "Small drones for community-based forest monitoring: An assessment of their feasibility and potential in tropical areas." *Forests* 5 (6): 1481–1507.

Peled, A., A. Bentur, and B. Mobasher. 2016. *Textile reinforced concrete*. London: Taylor & Francis.

Pereira, P. M. M., G. A. Monteiro, and D. M. F. Prazeres. 2015. "General aspects of biomimetic materials." In *Biotechnologies and biomimetics for civil engineering*, edited by F. P. Torgal, J. A. Labrincha, M. V. Diamanti, C.-P. Yu, and H. K. Lee, 57–79. Cham, Switzerland: Springer.

Portal, N., I. Fernandez, L. Thrane, and K. Lundgren. 2014. "Pull-out of textile reinforcement in concrete." *Constr. Build. Mater.* 71: 63–71.

Ranade, R., V. C. Li, W. F. Heard, and B. A. Williams. 2017. "Impact resistance of high strength-high ductility concrete." *Cem. Concr. Res.* 98: 24–35.

Reda Taha, M. M., and J. Lucero. 2005. "Damage identification for structural health monitoring using fuzzy pattern recognition." *Eng. Struct.* 27 (12): 1774–1783.

Reda Taha, M. M., A. Noureldin, N. El-Sheimy, and N. G. Shrive. 2004. "Feedforward neural networks for modelling time-dependent creep deformations in masonry structures." In *Proc., Inst. Civ. Eng. Struct. Build.* 157 (SB4): 279–292.

Restas, A. 2015. "Drone applications for supporting disaster management." *World J. Eng. Technol.* 3 (03): 316–321.

Ross, T. 2016. *Fuzzy logic with engineering applications*. 4th ed. Hoboken, NJ: Wiley.

Salehi, H., and R. Burgueño. 2018. "Emerging artificial intelligence methods in structural engineering." *Eng. Struct.* 171: 170–189.

Shahin, M. A. 2013. "Artificial intelligence in civil engineering: Applications, modeling aspects and future directions." *Metaheuristics Water Geotech. Transp. Eng.* 169–204. Accessed March 12, 2021. https://doi.org/10.1016/B978-0-12-398296-4.00008-8

Sherif, M. M., and O. E. Ozbulut. 2018. "Tensile and superelastic fatigue characterization of NiTi shape memory cables." *Smart Mater. Struct.* 27 (1): 015007.

Shin, D. H., and P. S. Dunston. 2009. "Evaluation of augmented reality in steel column inspection." *Autom. Constr.* 18 (2): 118–129.

Soga, K., and L. Luo. 2018. "Distributed fiber optics sensors for civil engineering infrastructure sensing." *J. Struct. Integrity Maint.* 3 (1): 1–21.

Sohn, H., J. A. Czarnecki, and C. R. Farrar. 2000. "Structural health monitoring using statistical process control." *J. Struct. Eng.* 126: 1356–1363.

Soliman, E. M., U. F. Kandil, and M. M. Reda Taha. 2012. "The significance of carbon nanotubes on styrene butadiene rubber (SBR) and SBR modified mortar." *Mater. Struct.* 45 (6): 803–816.

Speck, T., and N. P. Rowe. 2006. "How to become a successful climber—Mechanical, anatomical, ultra-structural and biochemical variations during ontogeny in plants with different climbing strategies." In Vol. 1 of *Proc. 5th Int. Plant Biomechanics Conf.*, 103–108.

Stent, S., R. Gherardi, B. Stenger, K. Soga, and R. Cipolla. 2016. "Visual change detection on tunnel linings." *Mach. Vision Appl.* 27 (3): 319–330.

Succar, B. 2009. "Building information modelling framework: A research and delivery foundation for industry stakeholders." *Autom. Constr.* 18 (3): 357–375.

Succar, B. 2010. "The five components of BIM performance measurement." In *CIB World Congress*.

Sun, Y. Y., Z. W. Yu, and Z. G. Wang. 2016. "Bioinspired design of building materials for blast and ballistic protection." *Adv. Civ. Eng.* 2016: 5840176.

Sutherland, I. E. 1968. "A head-mounted three-dimensional display." In *Proc. Fall Joint Computer Conf., Part I*, 757–764. New York: ACM.

Tazarv, M., and M. Saiidi. 2015. "Reinforcing NiTi superelastic SMA for concrete structures." *J. Struct. Eng.* 141 (8): 04014197.

Triantafillou, T., ed. 2016. *Textile fiber composites in civil engineering*. Oxford, UK: Elsevier Science.

Van Tittelboom, K., and N. De Belie. 2013. "Self-healing in cementitious materials—A review." *Materials* 6 (6): 2182–2217.

Volk, R., J. Stengel, and F. Schultmann. 2014. "Building information modeling (BIM) for existing buildings—Literature review and future needs." *Autom. Constr.* 38: 109–127.

Volkova, A., A. Paykov, S. Semenov, O. Stolyarov, and B. Melnikov. 2016. "Flexural behavior of textile-reinforced concrete." *MATEC Web Conf.* 53: 01016.

Webster, A., S. Feiner, B. MacIntyre, W. Massie, and T. Krueger. 1996. "Augmented reality in architectural construction, inspection and renovation." In *Proc., ASCE 3rd Congress in Computing in Civil Engineering*, 913–919. Reston, VA: ASCE.

Wegst, U. G., H. Bai, E. Saiz, A. P. Tomsia, and R. O. Ritchie. 2015. "Bioinspired structural materials." *Nat. Mater.* 14 (1): 23–36.

Wen, S., and D. D. L. Chung. 2002. "Piezoelectric cement-based materials with large coupling and voltage coefficients." *Cem. Concr. Res.* 32: 335–339.

White, S. R., N. R. Sottos, P. H. Geubelle, J. S. Moore, M. R. Kessler, S. R. Sriram, et al. 2001. "Autonomic healing of polymer composites." *Nature* 409 (6822): 794–797.

Williams Portal, N., K. Lundgren, H. Wallbaum, and K. Malaga. 2015. "Sustainable potential of textile-reinforced concrete." *J. Mater. Civ. Eng.* 27 (7): 04014207.

Wu, M., B. Johannesson, and M. Geiker. 2012. "A review: Self-healing in cementitious materials and engineered cementitious composite as a self-healing material." *Constr. Build. Mater.* 28 (1): 571–583.

Yang, Y., Z. Wei, C. Wang, and Z. Tong. 2013. "Versatile fabrication of nanocomposite microcapsules with controlled shell thickness and low permeability." *ACS Appl. Mater. Interfaces* 5: 2495–2502.

Yoon, H., J. Shin, and B. F. Spencer. 2018. "Structural displacement measurement using an unmanned aerial system." *Comput.-Aided Civ. Infrastruct. Eng.* 33 (3): 183–192.

Yu, K.-Q, J.-T. Yu, J.-G. Dai, Z.-D. Lu, and S. P. Shah. 2018. "Development of ultra-high performance engineered cementitious composites using polyethylene (PE) fibers." *Constr. Build. Mater.* 158: 217–227.

Yu, X., and E. Kwon. 2009. "A carbon nanotube/cement composite with piezoresistive properties." *Smart Mater. Struct* 18: 1–5.

Yuan, Y. C., M. Z. Rong, M. Q. Zhang, J. Chen, G. C. Yang, and X. M. Li. 2008. "Self-healing polymeric materials using epoxy/mercaptan as the healant." *Macromolecules* 41 (14): 5197–5202.

Žák, J., P. Štemberk, and J. Vodička. 2017. "Production of a textile reinforced concrete protective layers with non-woven polypropylene fabric." *IOP Conf. Ser.: Mater. Sci. Eng.* 246 (1): 012054.

Zastrau, B., M. Richter, and I. Lepenies. 2003. "On the analytical solution of pullout phenomena in textile reinforced concrete." *J. Eng. Mater. Technol.* 125 (1): 38–43.

Zhang, Y., Y. Zhang, G. Liu, Y. Yang, M. Wu, and B. Pang. 2018. "Fresh properties of a novel 3D printing concrete ink." *Constr. Build. Mater.* 174: 263–271.

Zhou, T., X. Wang, P. Cheng, T. Wang, D. Xiong, and X. Wang. 2013. "Improving the thermal conductivity of epoxy resin by the addition of a mixture of graphene nanoplatelets and silicon carbide microparticles." *Express Polym. Lett.* 7 (7): 584–594.

Ziolkowski, P., and M. Niedostatkiewicz. 2019. "Machine learning techniques in concrete mix design." *Materials* 12 (8): 1256.

APPENDIX
TERMINOLOGY

This appendix provides definitions of terminology, many adapted from Ayyub (2014a). Noteworthily, some terms have multiple definitions as a result of dealing with an emerging field with contemporary applications. This richness was retained without an attempt to resolve or reach consensus by the authors.

Adaptive risk management is defined as the coordinated and adaptive activities with regard to risk. See also *risk management*.

Additive manufacturing is the process of digital manufacturing of parts and structures incorporating the design and formation of three-dimensional computer graphic models using 3D scanning and computer-aided design technologies and production of three-dimensional products by 3D printing of metals, concrete, plastics, and potentially biological materials.

Agent-based modeling is a class of computational models for simulating actions and interactions of autonomous agents with a view to assess their effects on the system as a whole.

Artificial intelligence (AI) is a term used to describe enabling intelligent behavior in artifacts. This means enabling those artifacts to have perception, reasoning, knowledge representation, planning, learning, language processing, and the ability to manipulate information or objects in their environments (Lugar and Stubblefield 1993).

Asset management is the coordinated activity of an organization to realize value from assets.

Augmented reality is the process of integrating or complimenting real-world objects with virtual information (e.g., graphics) to enhance users' understanding and interaction with the surrounding environment.

When the real world is replaced by virtual objects, the process is known as virtual reality.

Building information modeling is a collaborative framework that utilizes computer modeling to provide an interactive platform that enables virtual design, construction, and occupation of the building and connect building information to owners, design engineers, construction managers and engineers, and contractors (Kensek and Noble 2014).

Computable general equilibrium model is a macroeconomic method of economic interdependencies within a regional economy. It is used to simulate a complete market economy in which prices and quantities supplied and demanded are adjusted to clear all markets.

Consequence is the immediate, short-term, and long-term effects of an event affecting objectives, for example, sea-level rise. These effects may include human and property losses, environmental damages, and loss of lifelines.

Countermeasure is an action taken or a physical capability provided with the principal purpose of reducing or eliminating vulnerabilities or reducing the occurrence of attacks.

Critical infrastructure consists of systems and assets, whether physical or virtual, which are vital to a nation that the incapacity or destruction of such systems and assets would have a debilitating impact on security, national economic security, national public health or safety, or any combination of these matters.

Design basis value is the accepted performance level for design that might include a factor of safety.

Discount rate (sometimes termed *hurdle rate, interest rate, cutoff rate, benchmark, or the cost of capital*) is an interest as a percentage assigned to capture the time value of money. Discounting is used to normalize future values into present values, facilitating comparisons of cash flows occurring at different times across the study period.

Exposure is the extent to which an organization's or stakeholder's concerns are subject to an event and defined by things at risk that might include population at risk, property at risk, and ecological and environmental concerns at risk.

FLEX approach is a collection of diverse and flexible coping strategies recommended by the Nuclear Energy Institute to provide in-depth resilience against a nuclear accident risk at a nuclear power plant facility.

Fragility curve for a system shows the probability of exceeding a particular damage state (or performance including failure) as a function of a demand parameter on the system that represents the hazard level, such as ground motion (e.g., spectral displacement at a given frequency).

Fragility is the conditional probability of failure at a particular hazard level. Fragility can be measured by the conditional probability of failure associated with the occurrence of a hazard of a particular intensity (i.e., the conditional probability that the associated required capacity exceeding the available capacity).

Hazard curve is a plot of the magnitude of a hazard type and the associated frequency of occurrence (this definition is strictly valid for only stationary processes); for a nonstationary process (e.g., due to impact of climate change), the hazard curve reflects projection based on the presently available data.

Hazard is a source of potential harm or a condition, which may result from an external cause (e.g., earthquake, flood, or human agency) or an internal cause, with the potential to initiate a failure mode.

Ignorance is deficiency in knowledge. Within the realm of conscience ignorance, incompleteness and inconsistency are the primary categories defining it.

Infrastructure is defined as the basic physical and organizational structures and facilities (e.g., building clusters and lifeline systems, needed for the operation of a society or enterprise).

Functionality (of infrastructure) is the quality or state of working properly to provide a regular reliable service at, or as close as possible to, what the infrastructure system provided prior to an event.

Operability (of infrastructure) is the fitness, capacity, or ability to use to provide basic services allowing customers/users to receive normal, or near normal, amenities from a potentially impaired infrastructure system following an event.

Infrastructure resilience analysis methodology is a hybrid resilience assessment approach that combines performance-based metrics with attribute-focused analysis.

Input–output model is a macroeconomic method that models economic interdependencies within a regional economy. It is used to simulate flows throughout a market economy. It does not allow for price adjustments.

Internal rate of return is the discount rate that results in the net present value equaling zero.

Life-cycle cost is a measure of all costs of a potential investment in life in present-value terms.

Likelihood is the chance of something happening, whether defined, measured, or determined objectively or subjectively, qualitatively, or quantitatively, and described using general terms or mathematically, such as a probability or a frequency over a given time period.

Machine learning is a term used to describe a subset of AI where data trends are learned and models are developed to predict the future performance of systems.

Maximum credible level is an estimate associated with a confidence level or percentile level such as 90%.

Maximum credible earthquake level ground motions is the largest earthquakes that can reasonably be expected.

Maximum tolerable period of disruption is the maximum duration that is deemed acceptable for interruption of a facility or network's operations.

Net present value is the current value of all benefits minus all costs.

Network efficiency or global network efficiency is a measure of how efficient communication or flow is between any two nodes in a network.

Network or graph, in its simplest form, is a collection of points joined together in pairs by lines. Points are referred to as vertices or nodes, and lines are termed links or edges.

Network resilience is the persistence of its functions and performances under uncertainty in the face of disturbances.

Network robustness is the ability of the system and system components to sustain external shocks without significant degradation of performance. The robustness of a network, therefore, is its ability to withstand residual node connectivity after a possible adverse event resulting in disruptions.

Network vulnerability is an internal susceptibility of a system to component failures that are exposed to external shocks.

Planning horizon is the study period of the analysis.

Present value is the current value of cash flow occurring over time.

Probability is a measure of the chance of occurrence, likelihood, odds, or degree of belief that a particular outcome or event will occur, expressed as a number between 0 and 1, where 0 is impossibility and 1 is absolute certainty. This measure meets the axioms of probability theory. Probability has at least two primary interpretations: (1) a frequency representing the occurrence fraction of an outcome in repeated trials or an experiment as sometimes termed an objective probability and (2) subjective probability that is based on the state of knowledge.

Probable maximum value is the central tendency of the maximum value.

Rapidity is the capacity of a system to reduce losses and meet performance requirements by recovering in a timely manner (Biringer et al. 2013).

Redundancy is the ability of a system to function even after some of its components or subsystems have failed or, more broadly, the extent to which the system satisfies and sustains functional requirements after the failure of some of its components or subsystems (Biringer et al. 2013).

Reliability is the ability of a system or component to meet target performance levels during a time period and under set conditions of interest.

Residual risk is the amount of risk remaining after realizing the net effect of risk-reducing actions.

Resilience is defined by the Presidential Policy Directive (PPD)-8 (PPD 2011) as "the ability to adapt to changing conditions and withstand and rapidly recover from disruption due to emergencies." PPD-21 (PPD 2013) expanded the definition to include "the ability to prepare for and adapt to changing conditions and to withstand and recover rapidly from disruptions. Resilience includes the ability to withstand and recover from deliberate attacks, accidents, or naturally occurring threats or incidents." The National Institute of Standards Technology Community Resilience Planning Guide for Buildings and Infrastructure Systems (NIST 2016) adopted these definitions. The resilience of a system's function can be measured based on the persistence of a corresponding functional performance under uncertainty in the face of disturbances (Ayyub 2014b). This definition is consistent with the ISO (2009) risk definition of the "effect of uncertainty on objectives."

Resilience-based Earthquake Design Rating System is a resilience-based earthquake design framework for owners, architects, and engineers to explore resilience enhancement opportunities for an existing facility as well as for design a new facility.

Resourcefulness is the capacity of a system to mobilize materials and/or human resources to recover and meet performance goals after a disruption event (Biringer et al. 2013). It is the ability to diagnose and prioritize problems and to mobilize solutions by identifying and monitoring all resources, including economic, technical, and social information.

Risk acceptance is the degree of risk associated with a system or endeavor that a decision maker perceives and accepts the associated actions under a given set of circumstances and with the associated costs. A decision maker's risk tolerance and resources are the foundation of risk acceptance.

Risk analysis is the technical and scientific process to comprehend the nature of risk and to determine the level of risk by examining the underlying components or elements of risk.

Risk assessment is an overall process of (1) risk identification, (2) risk analysis, and (3) risk evaluation.

Risk attitude is an organization's approach to assess and eventually pursue, retain, take, or turn away from risks.

Risk aversion is the attitude to turn away from risk.

Risk communication involves perceptions of risk and depends on the audience targeted; hence, it is classified into risk communication to the media, public, and engineering community.

Risk context is the external and internal parameters or considerations to be taken into account when managing risks and setting the scope and risk criteria for the risk management policy.

Risk criteria are the terms of reference against which the significance of a risk is evaluated, reflecting the organizational objectives expressed in external and internal contexts and in keeping with standards, laws, policies, and other requirements.

Risk identification is the process of finding, recognizing, and describing risks including sources, events, scenarios, and their causes and potential consequences involving historical data, theoretical analysis, informed and expert opinions, and stakeholders' needs.

Risk management is the coordinated activities to direct and control an organization with regard to risk following a framework consisting of designing, implementing, monitoring, reviewing, and continually improving risk management throughout the organization. Risk management should be founded in strategic and operational policies, objectives, mandates, practices, and commitments through organizational arrangements including plans, relationships, accountabilities, resources, processes, and activities.

Risk neutrality is having the same attitude regardless of the potential loss.

Risk owner is a person or entity with the accountability and authority to manage a risk.

Risk seeking is the attitude to pursue, retain, or undertake the risk for potential return.

Risk should be associated with a system and is the potential loss resulting from an uncertain exposure to a hazard or resulting from an uncertain event that exploits the system's vulnerability. Risk should be based on identified risk events or event scenarios. ISO (2009) provided a broadly applicable definition of risk in its standard as the "effect of uncertainty on objectives" in order to cover the following considerations as noted in the standard: (1) An effect is a deviation from the expected that can be positive and/or negative effect. (2) Objectives can have different aspects, such as financial, health and safety, and environmental goals, and can apply at different levels, such as strategic, organization-wide, project, product, and process. (3) Risk is often expressed in terms of a combination of the consequences of an event, including changes in circumstances, and the associated likelihood of occurrence as provided in the commonly used definition.

Risk tolerance is the degree of risk associated with normal daily activities that people tolerate, usually without making a conscious decision. As for organization or stakeholders, it is the readiness to bear the risk after risk treatment to achieve its objectives. Risk tolerance can be influenced by legal or regulatory requirements.

Robustness is the ability of elements or systems to withstand a significant level of performance demand without suffering loss of function.

Safety is a judgment related to risk tolerance, or acceptability in the case of decision making, for the system.

Savings-to-investment ratio is discounted savings divided by discounted investment costs.

Scenario is defined as joint events and system state that lead to an outcome of interest. A scenario defines a suite of circumstances of interest in a risk assessment. Thus, there may be loading scenarios, failure scenarios, or downstream flooding scenarios. A scenario can also be defined as the joint occurrence of events following a particular order or sequence in occurrence.

Smart materials have the ability to respond to external excitations differently based on the ability to recognize and differentiate between these excitations, based on their multiple functions such as self-healing, self-sensing, and so on.

Social accounting matrix is a matrix that defines the linkages of (1) households and producers through product and capital markets through savings and sales and (2) government, households, and producers through taxes and transfers.

Stakeholder is a person, such as a decision maker and owner, or an organization that can affect, be affected by, or perceive themselves to be affected by a decision or activity.

Structural health monitoring is using technology to monitor the structure's health. This term is used to describe the process that includes sensing of structure's response using different technologies, feature extraction of the sensed data to realize sensitive feature(s) that can describe the structure's health, patterning those features and relating those patterns to structure's behavior for damage pattern recognition and finally, using those patterns for structural prognosis to determine the time necessary for repair, maintenance, or replacement (Reda Taha et al. 2006).

Sustainability, according to ASCE (2016) in its Policy Statement 418, is a set of economic, environmental, and social conditions in which all of the society has the capacity and opportunity to maintain and improve its quality of life indefinitely, without degrading the quantity, quality, or the availability of natural, economic, and social resources. Sustainable development is the application of these resources to enhance the safety, welfare, and quality of life for all of society. Several other definitions are available as provided by Webb and Ayyub (2016): (1) "Creating and maintaining conditions under which humans and nature can exist in productive harmony and that permit fulfilling social, economic, and other requirements of present and future generations" (EPA 2016a). (2) "Ability to maintain or improve standards of living without damaging or depleting natural resources for present and future generations" (EPA 2016b). (3) "The creation of manufactured products that use processes that minimize negative environmental impacts, conserve energy and natural resources, are safe for employees, communities, and consumers and are economically sound" (ITA 2016). (4) "The practice of increasing the efficiency with which buildings and their sites use and harvest energy, water, and materials; and protecting and restoring human health and the environment, throughout the building life-cycle: siting, design, construction, operation, maintenance, renovation and deconstruction" (EPA 2016c).

System is a group of interacting, interrelated, or interdependent elements, such as people, property, materials, environment, and processes.

System of systems (SoS) is an assemblage of components that individually may be regarded as systems and that possess two additional properties: operational independence of components and managerial independence of the components.

Uncertainty is the state of deficiency in information with two uncertainty types identified as follows: (1) Aleatory uncertainty is the inherent, random, or nonreducible uncertainty, such as material strength randomness. (2) Epistemic uncertainty is the knowledge-based, subjective uncertainty that can be reduced with the collection of data or attainment of additional knowledge.

Vulnerability is defined as the intrinsic properties of a system making it susceptible to a hazard or a threat or a risk source that can lead to an event with a consequence or is an inherent state of the system, for example, physical, technical, organizational, or cultural, that can be exploited by an adversary to cause harm or damage.

REFERENCES

ASCE. 2016. *Policy statement 418*. Reston, VA: ASCE. Accessed on February 9, 2016. http://www.asce.org/issues-and-advocacy/public-policy/policy-statement-418—the-role-of-the-civil-engineer-in-sustainable-development/.

Ayyub, B. M. 2014a. *Risk analysis in engineering and economics*. 2nd ed. Boca Raton, FL: Chapman & Hall/CRC Press.

Ayyub, B. M. 2014b. "Systems resilience for multihazard environments: Definition, metrics, and valuation for decision making." *J. Risk Anal.* 34 (2): 340–355.

Biringer, B. E., E. D. Vugrin, and D. E. Warren. 2013. *Critical infrastructure system security and resiliency*. Boca Raton, FL: CRC Press.

EPA. 2016a. "Learn about sustainability." Accessed February 9, 2016. http://www.epa.gov/sustainability/learn-about-sustainability.

EPA. 2016b. "EPA's report on the environment." Accessed February 9, 2016. http://cfpub.epa.gov/roe/chapter/sustain/index.cfm.

EPA. 2016c. "Green buildings." Accessed February 9, 2016. http://archive.epa.gov/greenbuilding/web/html/faqs.html.

ISO. 2009. Risk Management—Principles and Guidelines, ISO 31000, iso.org, Geneva.

ITA (International Trade Administration), Department of Commerce. 2016. "How does commerce define sustainable manufacturing?" Accessed February 9, 2016. http://www.trade.gov/competitiveness/sustainable-manufacturing/how_doc_defines_SM.asp.

Kensek, K., and D. Noble. 2014. *Building information modeling: BIM in current and future practice*. Hoboken, NJ: Wiley.

Stubblefield, W. A., and G.F. Luger. 1993. *Artificial intelligence: structures and strategies for complex problem-solving*, 2nd ed. Redwood City, CA: Benjamin-Cummings Publishing.

NIST. 2016. Vol. 1 of *Community resilience planning guide for buildings and infrastructure systems*. NIST Special Publication 1190. Gaithersburg, MD: National Institute of Standards and Technology.

PPD (Presidential Policy Directives). 2011. "National preparedness." PPD-8. Accessed February 9, 2016. http://www.dhs.gov/presidential-policy-directive-8-national-preparedness.

PPD. 2013. "Critical infrastructure security and resilience." PPD-21. Accessed February 9, 2016. https://www.whitehouse.gov/the-press-office/2013/02/12/presidential-policy-directive-critical-infrastructure-security-and-resil.

Reda Taha, M. M., A. Noureldin, J. L. Lucero, and T. J. Baca. 2006. "Wavelet transform for structural health monitoring: A compendium of uses and features." *Struct. Health Monit.* 5 (3): 267–295.

Webb, D., and B. M. Ayyub. 2016. "Sustainable construction and manufacturing. 1: Definitions, metrics, and valuations for decision making." *ASCE-ASME J. Risk Uncertainty Eng. Syst. Part A: Civ. Eng.* 3(3), Accessed March 12, 2021. https://doi.org/10.1061/AJRUA6.0000893

INDEX

Note: Page numbers followed by *f* and *t* indicate figures and tables.